Lecture Notes in Computer Science 14918

The series Lecture Notes in Computer Science (LNCS), including its subseries Lecture Notes in Artificial Intelligence (LNAI) and Lecture Notes in Bioinformatics (LNBI), has established itself as a medium for the publication of new developments in computer science and information technology research, teaching, and education.

LNCS enjoys close cooperation with the computer science R & D community, the series counts many renowned academics among its volume editors and paper authors, and collaborates with prestigious societies. Its mission is to serve this international community by providing an invaluable service, mainly focused on the publication of conference and workshop proceedings and postproceedings. LNCS commenced publication in 1973.

Joe Tekli · Johann Gamper · Richard Chbeir ·
Yannis Manolopoulos
Editors

Advances in Databases and Information Systems

28th European Conference, ADBIS 2024
Bayonne, France, August 28–31, 2024
Proceedings

Springer

Editors
Joe Tekli
Lebanese American University Engineering School
Lebanese American University
Chouran Beirut, Lebanon

Richard Chbeir
Université de Pau et des Pays de l'Adour
Anglet, France

Johann Gamper
Free University of Bozen-Bolzano
Bozen-Bolzano, Italy

Yannis Manolopoulos
Open University of Cyprus
Nicosia, Cyprus

ISSN 0302-9743 ISSN 1611-3349 (electronic)
Lecture Notes in Computer Science
ISBN 978-3-031-70628-8 ISBN 978-3-031-70626-4 (eBook)
https://doi.org/10.1007/978-3-031-70626-4

This Springer imprint is published by the registered company Springer Nature Switzerland AG
The registered company address is: Gewerbestrasse 11, 6330 Cham, Switzerland

Preface

This volume contains the main track papers presented at the 28th European Conference on Advances in Databases and Information Systems (ADBIS 2024), held August 28–31, 2024, in Bayonne, France.

The first ADBIS Conference was held in Saint Petersburg, Russia (1997). Since then, ADBIS has been continuously organized as an annual event. Its previous editions were held in: Barcelona, Spain (2023); Torino, Italy (2022); Tartu, Estonia (2021); Lyon, France (2020); Bled, Slovenia (2019); Budapest, Hungary (2018); Nicosia, Cyprus (2017); Prague, Czech Republic (2016); Poitiers, France (2015); Ohrid, North Macedonia (2014); Genoa, Italy (2013); Poznań, Poland (2012); Vienna, Austria (2011); Novi Sad, Serbia (2010); Riga, Latvia (2009); Pori, Finland (2008); Varna, Bulgaria (2007); Thessaloniki, Greece (2006); Tallinn, Estonia (2005); Budapest, Hungary (2004); Dresden, Germany (2003); Bratislava, Slovakia (2002); Vilnius, Lithuania (2001); Prague, Czech Republic (2000); Maribor, Slovenia (1999); and Poznań, Poland (1998). The official ADBIS portal at http://adbis.eu provides up-to-date information on all ADBIS Conferences, persons in charge, publications, and issues related to the ADBIS community.

The program of ADBIS 2024 included keynote talks, research papers, tutorials, workshops, a doctoral consortium (DC), as well as a diversity, equity, and inclusion (DEI) panel and keynote. The main conference attracted 43 paper submissions. All the papers went through a process of rigorous single-blind reviewing by at least three reviewers. Eventually, the Program Committee selected 15 submissions as full contributions, appearing in this Springer LNCS proceedings volume, and 7 short papers, appearing in Springer's Communications in Computer and Information Science (CCIS) series. The 15 papers accepted as full contributions amount to an acceptance rate of 35%.

The selected papers span a large spectrum of topics in the broader field of data management. We have organized the papers in seven sessions: (1) Algebra, Models, and Schemata; (2) Discovery and Data Analysis; (3) Algorithms and Optimization; (4) Access Methods and Query Processing; (5) Advanced Architectures, (6) Machine Learning, and (7) Large Language Models.

Following a successful tradition, selected best papers of ADBIS 2024 will be invited for a special issue of the following journals: Information Systems (Springer), SN Computer Science (Springer), Digital (MDPI), and Computer Science & Information Systems. Therefore, the PC chairs would like to express their sincere gratitude to the Editors-in-Chief of the above journals for their approval regarding these special issues.

For this ADBIS event we had keynote talks by experts in different fields of the broader area of data management. In particular (in alphabetical order):

- Angela Bonifati (CNRS Liris research lab, Lyon 1 University, France) delivered a talk on "*The Future is Hybrid Graphs: Combining Graphs with Time Series*".

- Christian S. Jensen (Center for Data Intensive Systems, Aalborg University, Denmark) dealt with the topic of *"Temporal and Spatio-Temporal Data Management and Analytics – A Personal Perspective"*, and
- Schahram Dustdar (TU Wien, Austria) covered the topic of *"Distributed Intelligence in the Computing Continuum"*.

In addition, two excellent tutorials were also included in the program, which were delivered by:

- Zheying Zhang and Kostas Stefanidis (Tampere University, Finland) who presented a tutorial entitled *"Data-Driven Analysis for Monitoring Software Evolution"*, and
- Witold Andrzejewski, Bartosz Bębel, Paweł Boiński and Robert Wrembel (Poznań University of Technology, Poland) who delivered a tutorial entitled *"On Customer Data Deduplication - Research vs. Industrial Perspective: Lessons Learned from an R&D Project in the Financial Sector"*.

The ADBIS 2024 conference hosted four affiliated workshops:

- 6th Workshop on Modern Approaches in Data Engineering and Information System Design (MADEISD),
- 5th Workshop on Intelligent Data - From Data to Knowledge (DOING),
- 3rd Workshop on Personalization and Recommender Systems (PERS), and
- 3rd Workshop on Knowledge Graphs Analysis on a Large Scale (K-GALS).

All four workshops addressed cutting-edge topics that call for extensive research from both a foundational and an application perspective to progress the state of the art. In total, 37 papers were submitted to these workshops, out of which 17 were selected for presentation and publication, yielding an overall acceptance rate of 46%. The accepted workshop papers are published in the accompanying proceedings volume published in the CCIS series by Springer. This volume also contains the contributions of the Doctoral Consortium stream; the latter call attracted 20 submissions, out of which 11 were selected for inclusion in the program.

The ADBIS 2024 conference prioritized the DEI initiative throughout the experience. Authors and participants were encouraged to follow DEI-aware communication guidelines when preparing papers and presentations to ensure inclusive communication. We gained valuable insights into our community's DEI needs through a pre-conference anonymous survey. We actively supported DEI at the conference by collaborating with the organizing committee to offer two honorary registration grants specifically for scientists participating in the DC and DEI sessions. This helped to ensure broader participation and foster a more inclusive research environment. The ADBIS conference program featured three DEI plenary sessions. Additionally, a data analytics activity with the DC Chairs gave Ph.D. students practical skills for designing equity-aware data-driven experiments. The DEI stream offered a keynote address by:

- Rita Bencivenga (University of Genoa, Italy), who delivered a talk on *Incorporating Gender+ and Intersectionality into Research Practices.*

Furthermore, a panel discussion entitled "*More Than Muscles in the Lab: Expanding Masculinities in a Competitive Research Culture*" brought together distinguished researchers who study masculinity in the scientific community, exploring its various perspectives and impact. Our distinguished panelists were Rita Bencivenga (University of Genoa), Jorge Alberto Hidalgo Toledo (Anáhuac University, Mexico), Amir Hossein Payberah (KTH Royal Institute of Technology, Sweden), Arun Kumar (University of California, San Diego, USA) and Romain Sabathier (COOP-EGAL, France).

We want to thank everyone who contributed to the ADBIS 2024 conference. The list is long, and we resort to grouping our thanks to the following groups of persons:

- Authors, without their hard work the conference would not be possible.
- Keynote speakers, Tutorial presenters, and panelists for sharing their wisdom with the conference audience.
- Members of the Program Committee and External Reviewers, for all the hard work in diligently reviewing, commenting, and discussing the submitted papers and for their useful suggestions to the authors.
- Members of the ADBIS Steering Committee for their trust and support.
- Members of the Organizing Committee for taking care of several different tasks related to local affairs, without whom the conference would have not been such a pleasant and inspiring experience.
- The CMT staff for hosting the process of paper submission, reviewing, and collection of the final material.
- Springer staff for making it possible to publish the proceedings of ADBIS 2024 conference in the LNCS and CCIS series.

Thanks are due also to the sponsors of the ADBIS 2024 conference, namely OpenCEMS, Université de Pau et des Pays de l'Adour, IUT de Bayonne, and Springer.

July 2024

Joe Tekli
Johann Gamper
Richard Chbeir
Yannis Manolopoulos

Organization

General Chairs

Richard Chbeir	Université de Pau et des Pays de l'Adour, France
Yannis Manolopoulos	Open University of Cyprus, Cyprus

Program Co-chairs

Joe Tekli	Lebanese American University, Lebanon
Johann Gamper	Free University of Bozen-Bolzano, Italy

Workshops Chair

Salma Sassi	Université de Pau et des Pays de l'Adour, France

Special Issues Co-chairs

Dimitrios Katsaros	University of Thessaly, Greece
Oscar Romero	Universitat Politècnica de Catalunya, Spain

Doctoral Consortium Co-chairs

Ester Zumpano	Università della Calabria, Italy
Genoveva Vargas Solar	CNRS, France
Mirjana Ivanovic	University of Novi Sad, Serbia

Tutorials Co-chairs

Claudio Silvestri	Università Ca' Foscari Venezia, Italy
Goce Trajcevski	Iowa State University, USA

Diversity and Inclusion Co-chairs

Barbara Catania	University of Genoa, Italy
Genoveva Vargas Solar	CNRS, France

Publicity Chairs

Khouloud Salameh	AURAK, UAE
Theodoros Tzouramanis	University of Thessaly, Greece

Proceedings Chair

Karam Bou-Chaaya	Expleo Group, France

Webmasters

Elie Chicha	Université de Pau et des Pays de l'Adour, France
Fouad Achkouty	Université de Pau et des Pays de l'Adour, France

International Program Committee Members

Agata Filipowska	Poznań University of Economics, Poland
Alberto Abello	Universitat Politècnica de Catalunya, Spain
Alberto Gutiérrez Torre	Barcelona Supercomputing Center, Spain
Alejandro Maté	University of Alicante, Spain
Aleksandra Karpus	Gdańsk University of Technology, Poland
Alexandre Chanson	University of Tours, France
Alexandros Karakasidis	University of Macedonia, Greece
Andreas Behrend	University of Bonn, Germany
Andrzej Kozik	University of Opole, Poland
Angelo Montanari	University of Udine, Italy
Anton Dignos Libera	Università di Bolzano, Italy
Antonio Corral	University of Almería, Spain
Arpita Dutta	National University of Singapore, Singapore
Audrone Lupeikiene	Vilnius University, Lithuania
Bentayeb Fadila	Lyon University, France
Bernd Amann	Sorbonne Université, France

Bernhard Thalheim	Christian-Albrechts-Universität zu Kiel, Germany
Besim Bilalli	Universitat Politècnica de Catalunya, Spain
Boris Novikov	National Research University, Russia
Boussaid Omar	Université Lyon 2, France
Carmem S. Hara	Universidade Federal do Paraná, Brazil
Chalachew Muluken Liyew	University of Turin, Italy
Cheikh Ba	Université Gaston Berger, Senegal
Christos Doulkeridis	University of Piraeus, Greece
Claudia Steinberger	Universität Klagenfurt, Austria
Costin Badica	University of Craiova, Romania
Cristina D. Aguiar	Universidade de São Paulo, Brazil
Damiano Carra	University of Verona, Italy
Daniele Apiletti	Politecnico di Torino, Italy
Dejan Lavbic	University of Ljubljana, Slovenia
Dijana Oreški	University of Zagreb, Croatia
Drazen Brdjanin	University of Banja Luka, Bosnia and Herzegovina
Flavio Ferrarotti	Software Competence Centre Hagenberg, Austria
Francesco Guerra	Università di Modena e Reggio Emilia, Italy
Franck Ravat	Université de Toulouse, France
George Papastefanatos	ATHENA Research Center, Greece
Georgia Koloniari	University of Macedonia, Greece
Georgios Evangelidis	University of Macedonia, Greece
Giuseppe Polese	University of Salerno, Italy
Goran Velinov	Ss. Cyril and Methodius University in Skopje, North Macedonia
Gunter Saake	University of Magdeburg, Germany
Haridimos Kondylakis	Institute of Computer Science, CERTH, Greece
Isabelle Comyn-Wattiau	ESSEC Business School, France
Ivan Lukovic	University of Belgrade, Serbia
Jaroslav Pokorny	Charles University in Prague, Czechia
Javier A. Espinosa-Oviedo	University of Lyon, France
Jérôme Darmont	Université Lyon 2, France
Joe Tekli	Lebanese American University, Lebanon
Johann Eder	Alpen-Adria-Universität Klagenfurt, Austria
José R. R. Viqueira	University of Santiago de Compostela, Spain
Josep Berral	Barcelona Supercomputing Center, Spain
Julius Köpke	Universität Klagenfurt, Austria
Kemal Polat	Abant Izzet Baysal University, Turkey
Khalid Belhajjame	Université Paris-Dauphine, France
Kjetil Nørvåg	Norwegian University of Science and Technology, Norway

Ladjel Bellatreche	University of Poitiers, France
Lorena Etcheverry	Universidad de la República, Uruguay
Marcos Sfair Sunye	Universidade Federal do Paraná, Brazil
María del Mar Roldán	Universidad de Málaga, Spain
Mariella Bonomo	Università degli Studi di Palermo, Italy
Markus Endres	Munich University of Applied Sciences, Germany
Matteo Francia	University of Bologna, Italy
Miguel A. Martinez-Prieto	University of Valladolid, Spain
Milos Radovanovic	University of Novi Sad, Serbia
Milos Savic	University of Novi Sad, Serbia
Mirian Halfeld-Ferrari	Université d'Orléans, France
Moharram Challenger	University of Antwerp, Belgium
Mounir Zrigui	University of Monastir, Tunisia
Nicolas Labroche	Université de Tours, France
Nikolaos Polatidis	University of Brighton, UK
Panos Vassiliadis	University of Ioannina, Greece
Patrick Marcel	University of Orléans, France
Pawel Boinski	Poznań University of Technology, Poland
Petr Hajek	University of Pardubice, Czechia
Richard Chbeir	Université de Pau et des Pays de l'Adour, France
Robert Wrembel	Poznań University of Technology, Poland
Rogério Luís De Carvalho Costa	Polytechnic Institute of Leiria, Portugal
Sandro Bimonte	INRAE, France
Sara Migliorini	Università degli Studi di Verona, Italy
Sergey Stupnikov	Russian Academy of Sciences, Russia
Sergi Nadal	Universitat Politècnica de Catalunya, Spain
Simona E. Rombo	Università degli Studi di Palermo, Italy
Slavica Kordic	University of Novi Sad, Serbia
Sonja Ristic	University of Novi Sad, Serbia
Spiros Skiadopoulos	University of the Peloponnese, Greece
Stefan Jablonski	University of Bayreuth, Germany
Stefano Rizzi	University of Bologna, Italy
Sylvio Barbon	University of Trieste, Italy
Tadeusz Morzy	Poznań University of Technology, Poland
Tania Cerquitelli	Politecnico di Torino, Italy
Veronika Peralta	University of Tours, France
William Steingartner	Technical University of Košice, Slovakia
Witold Andrzejewski	Poznań University of Technology, Poland
Yannis Manolopoulos	Open University of Cyprus, Cyprus
Ylenia Galluzzo	Università degli Studi di Palermo, Italy
Yves Rybarczyk	Dalarna University, Sweden
Zoubida Kedad	University of Versailles, France

Contents

Algebra, Models, Schemata

JSON Model: a Lightweight Featureful DSL for JSON

Fabien Coelho(✉) and Claire Yannou-Medrala

Centre de recherche en informatique, Mines Paris – PSL, Paris, France
{fabien.coelho,claire.yannou-medrala}@minesparis.psl.eu

Abstract. JSON is a *de facto* standard cross-language textual format used to represent, exchange and store structured data. Data schemas allow to document structures and verify values. Three JSON-based schema description languages have been proposed for JSON so far: JSON Schema, JSound and JSound-C. These languages are verbose and use a lax validation semantics: 60% of public schemas have been found defective because JSON Schema is particularly error-prone. Based on these evidences, we present JSON Model, an alternative to previous proposals which is both lightweight and featureful, discuss key design choices, its efficient implementation, and its integration into OpenAPI.

Keywords: JSON · JSON Model · Schema description language · DSL · OpenAPI

1 Introduction

JSON [10] *JavaScript Object Notation* has become an ubiquitous cross-language *de facto* standard to represent, exchange and store data between computer applications. Its success stems from the extensive use of JavaScript [12] in web applications, and that many services rely on this technology. Like the more verbose XML, JSON can be parsed without knowing in advance the expected structure. It allows to serialize in textual form simple data structures built upon the *null* value, booleans, numbers, Unicode strings, arrays (list, tuple, sequence, set) and objects (struct, record, dict, map, association, key-value pairs), as illustrated in Fig. 1 with a simple object. It has a limited number of types, no syntax for comments, unbounded but incomplete numbers (e.g. no *NaN*). Object properties (key, attribute, field) are restricted to strings. Only tree structures can be serialized: There is no sharing of values or cycle. Thanks to these simple features, a wide range of libraries and tools are available for many programming languages and systems beyond JavaScript including Python, Java, Shell, and SQL.

J. Tekli et al. (Eds.): ADBIS 2024, LNCS 14918, pp. 3–17, 2024.
https://doi.org/10.1007/978-3-031-70626-4_1

```
{ "name": "Susie", "age": 6, "friends": [ "Calvin", "Hobbes" ] }
```

Fig. 1. A JSON object with 3 properties

JSON data are mostly generated and processed from a programming language: Humans prefer unquoted structured languages, e.g. YAML [15], to write structured files such as configurations, or lightweight markup languages such as Markdown [14] for text formatting. JSON has two main overlapping use cases:

API Data for simple structures exchanged at API interfaces, for instance between web front ends and back ends in multitier architectures;

Documents for possibly large loosely-structured textual data which are stored, transmitted, processed and finally displayed for direct human consumption.

The overlap comes from development practices based on JavaScript dynamic typing (lack of) discipline and loose document-oriented schema-less databases such as MongoDB which do not require data schemas to be formally declared. Another example of overlapping usage is open data documentations: They are structured data often accessed through an API, but they also need extensive meta-data for documentation purpose.

After this introduction, Sect. 2 discusses existing data-structure description languages, with a focus on JSON: JSON Schema is verbose and error-prone resulting in a majority of public schemas to be defective, and other languages are also verbose or less expressive. Section 3 presents JSON Model, a simpler, compact and expressive alternative, explaining its design choices, syntax and semantics. Section 4 introduces our proof-of-concept implementation, including preprocessing and compiler optimizations. Section 5 outlines its integration into OpenAPI, before concluding.

2 Related Work

As data structures are a key component of programming languages, describing them with various degrees of constraints is typically included in language syntaxes. This also applies to dynamically typed languages, e.g. TypeScript [5] has been developed to allow type declarations with JavaScript. These descriptions can be either tight or loose: tight does not allow much freedom about properties or types, whereas loose allows unknown properties. When considering cross-language features such as data interchange, language-independent description languages can been used, for instance SQL, Newgen [16] or CORBA IDL [6]. These description languages are tight, need specialized tools to manipulate actual data, and require to learn a syntax. Another approach is to use a file format such as XML and JSON which can be parsed without knowing in advance its structure, and to describe data separately, e.g. with a DTD or XSD for XML, and with JSON Schema, JSound or JSound-C for JSON. These languages are often loose by default.

JSON Schema [19] has been under development for 10 versions over nearly 15 years. The latest proposal (2020–12) defines 60 keywords to describe JSON

```
{ "$schema": "https://json-schema.org/draft/2020-12/schema",
  "title": "A Person",
  "type": "object",
  "properties": { "name": { "type": "string", "minLength": 1 },
                  "age": { "type": "integer", "minimum": 0 },
                  "friends": { "title": "The Person's Friends",
                               "type": "array",
                               "items": { "type": "string", "minLength": 1 },
                               "minItems": 1 } },
  "required": [ "name", "age" ],
  "additionalProperties": false }
```

Fig. 2. Tight JSON Schema for Fig. 1

structures, much like classic data structures, but also includes constraints on element sizes (strings, arrays...), uniqueness properties, regular expressions on property names, schema composition (`allOf anyOf oneOf`) and logical assertions (`if then else not`). Its open-document mindset means that schemas are loose by default, allowing any type or property unless otherwise stated. There is a meta-schema which can validate itself. Figure 2 tight schema is suitable for Fig. 1 data and is 4 times larger than this sample. JSON Schema complexity [2,4] induces many user errors: a majority of public schemas are found defective [20].

```
{ "types": [
    { "name": "Person",
      "kind": "object",
      "content": [ { "name": "name", "type": "string", "required": true },
                   { "name": "age", "type": "integer", "required": true },
                   { "name": "friends",
                     "type": { "name": "list-of-friends",
                               "kind": "array",
                               "content": "string" } } ],
      "closed": true } ] }
```

Fig. 3. Tight JSound for Fig. 1

```
{ "Person": { "!name": "string",
              "!age": "integer",
              "friends": [ "string" ] } }
```

Fig. 4. Loose JSound-C for Fig. 1

JSound [1] is an alternative JSON-based description format proposed to describe JSON and TYSON (a typed extension of JSON) data structures (Fig. 3). It is inspired by XML Schema and simpler than JSON Schema. It addresses some of its shortcomings, e.g. types are mandatory, and extensive types useful to programming languages are available. Although some JSON Schema features (composition operators, property name constraints) are not available, it provides inheritance. It shares the same open-document mindset with optional properties by default. JSound comes with the JSound-C compact variant (Fig. 4), which cannot be mixed with full JSound: the whole syntax must be switched to full JSound if any feature is not available, such as tight object definitions.

3 JSON Model

This section introduces an alternate JSON system for describing JSON data which is more compact than JSound-C, but still retains most features from JSON Schema. We present the design criteria, the basic syntax and semantics, the representation of values (constants, references, regex), constraints, composition and complements.

3.1 Design Criteria

JSON data structures exchanged through REST API by web or mobile applications are typically not-deeply-nested tight objects with few properties. We set 12 criteria to guide the conception of our powerful, structured and practical JSON description language.

- Four criteria define the power of the language, i.e. what can be described:
  - ***expressive*** power of description comparable to JSON Schema.
  - ***modular*** allow to import and reuse models if needed.
  - ***efficient*** allow fast (compiled) validation functions.
  - ***self-hosted*** simple JSON meta model.
- Four other criteria focus on its structure, i.e. how it will look and feel:
  - ***compact*** much shorter than JSON Schema and JSound.
  - ***composable*** by defining and referencing local models.
  - ***frugal*** avoid multiple redundant concepts.
  - ***universal*** human and computer language independent using symbols.
- Four drive the user experience, i.e. how practical it is:
  - ***intuitive*** easy to understand for humans, with usual operators.
  - ***discernible*** description easy to separate from what is described.
  - ***tight*** models should be tight by default, but easy to loosen.
  - ***convenient*** optimized for the simple data structure API use case.

3.2 Model Basics: Type Inference

In order to have a *compact* and *intuitive* representation, JSON Model relies on type inference [11], so that each type is represented by a value. The 5 basic types and 2 constructs of JSON are thus expressed directly: `null` as a model stands for the `null` value, `true` is a boolean value (`true` or `false`), `-1 0 1` are integers (any, non-negative and positive), `-1.0 0.0 1.0` are real numbers,

`""` (empty string) is a string, `[]` and `{}` are empty arrays and objects. The *convenient* definition of positive values stems from the observation that in most cases numbers are really expected to be positive, even if this is rarely declared.

Arrays are used to hold extended homogeneous lists or fixed tuples of heterogeneous items, thus it is *convenient* to prioritize these two usage patterns. A list is expressed as an array of one model, e.g. `[ 1 ]` is a list of strictly positive integers. A tuple is expressed as an array of several models: `[ "", "", [ 1.0 ] ]` is a tuple containing 3 items (a string, an other string, and a list of positive numbers). The induced non uniformity is worth for having the very frequent homogeneous list use case easily integrated. Moreover, as objects are often preferred over tuples, the later are rare.

Objects are the most versatile JSON constructs, possibly with mandatory, optional, and unknown properties. Following type inference principles, JSON Model expresses properties as properties, each property name mapped to its model. The first character of the name is used to express whether the property is mandatory (`!`, the default) or optional (`?`). Optional properties can also be described with references and regular expressions, as discussed in the next section. Objects are *tight* by default and can be loosen with the special `""` catch-all property, which is consistent with its interpretation as any string value. Mandatory and optional constant properties are directly validated. For other properties, first all optional references are tried: if the property name matches then the corresponding value must match as well. If properties did not match, all optional regex are tried. Finally, the catch-all property is applied.

```
{ "name": "", "age": 0, "?friends": [ "" ] }
```

Fig. 5. Tight JSON model for Fig. 1

Figure 5 shows a tight model for the Fig. 1 object with 3 properties. This tight model is slightly shorter than, and quite similar to, the sample value.

3.3 Model Constants and Types

Constants from different types are often needed in data structures, e.g. for declaring an enumeration or a tag value. In JSON Model, non-empty constant strings starting with an alphabetical character represent themselves, thus `"Susie"` means constant string `"Susie"`. All non-alphabetical first characters are reserved as a generic mechanism for extensions. Characters `?` `!` are used for property names (Sect. 3.2), and `_` `=` `$` `/` are defined below. Character `_` (underscore) is the escape prefix, thus `"_Susie"` is also constant string `"Susie"`, `"_|"` is simple string `"|"`, and `"_"` is the empty string, distinct from `""` which means any string. Other basic constants are introduced by the *intuitive* `=` (equal) followed by the expected value: `"=null"` is the `null` value, `"=false"` is boolean `false`, `"=42"` is integer `42`, and `"=6.02E23"` is the Avogadro number.

Character `$` (dollar) allows to reference predefined or locally defined models, *intuitive*ly similar to shell variables. As a convention, predefined values are in upper case. They include `$ANY` for any possible value, `$NONE` for no possible value, `$U32`, `$U64`, `$I32`, `$I64`, `$F32` and `$F64` for unsigned/signed integers and floats in their 32-bit and 64-bit variants, special string values such as `$REGEX` for regular expressions, `$URI` for URIs, `$DATE` for dates, `$UUID` for UUIDs... In some use cases, e.g. YAML integration (Sect. 5), it is *convenient* that all basic types have equivalent explicit names: `$STRING` for `""`, `$BOOL` for `true`, `$INTEGER` for `-1`, `$NUMBER` for `-1.0`, `$NULL` for `null`. More predefs can be added to express more types or constraints. References can also be used for constraining optional property names, provided that the value is a string.

A *convenient* use-case is to use string patterns with regular expressions. Character / (slash) introduces a regex which must end with /, like in Perl or JS. It can also be followed by options, e.g. `"/^[a-z]*$/i"` means all case-insensitive words in the Latin alphabet. Only safe (*efficient*) regular expressions are allowed by default, excluding backtracking implementations [17]. Implementations may optionally allow unsafe regex. Regex can be used as optional property names, matching any property of that name if it was not already matched.

```
{ "hero": "/^(Calvin|Susie)$/",        "?birth": "$DATE",
  "pi": "=3.14159274101257325",        "$URI": "$STRING",
  "forty-two": "=42",                  "/^(dai|week)ly$/": "$BOOL",
  "empty": "_",                        "": "$ANY"                    }
```

Fig. 6. Constants and Types

Figure 6 presents a model for an object with 4 mandatory properties: Property `hero` value matches a regex, `pi` is float π, `forty-two` is integer 42, `empty` is the empty string. Optional properties are allowed with constants, references and regex: `birth` is a valid date, URI-like properties are strings, `daily` and `weekly` are booleans, and any other property is valid with any value (loose object).

3.4 Model Constraints

JSON Schema has a dedicated syntax to validate constraints on values or sizes. In JSON Model, constraints are represented as an object with Property `@` (at sign) to denote the target model, and a limited number of additional properties to describe constraints. *Universal* symbolic properties `>=` `>` `<=` `<` `=` `!=` express length or order constraints, their interpretation depending on the values and types: For an array, an object or a string and if the value is an integer, the constraint is on the array size, number of properties or string length respectively; If both target type and value are strings, the constraint is a string lexicographic comparison. Optional constraints can be declared with additional keywords, e.g. `!` tells whether values are to be distinct in an array, or characters are distinct in a string. Unknown or inapplicable properties are rejected.

In the object described in Fig. 7: `login` has 8 to 10 lowercase characters, `birth` is at least in the XX-th century, `skills` are non-empty and distinct.

```
{ "login": { "@": "/^[a-z]*$/", ">=": 8, "<=": 10 },
  "birth": { "@": "$DATE", ">=": "1901-01-01" },
  "skills": { "@": [ "" ], ">": 0, "!": true } }
```

Fig. 7. Object with constrained properties using `@`

3.5 Model Composition

With these building blocks, data structures can be constructed from basic types in objects and arrays. To be more *expressive*, the next step is to combine elements with operators to indicate accumulation (**and**), alternative (**or**, **exclusive-or**) and object property merging (**merge**). JSON Model four *intuitive* combinators use objects with specific properties, one at a time, applied to an array of models. The one-at-a-time restriction simplifies the semantics and intelligibility of composed objects and its implementation is *efficient*.

The **or** combinator uses Property `|` (pipe) to express a union or tagged-union, or to list the values of an enumeration. The generalized **exclusive-or** (**xor**), with Property `^` (caret), is a particular case where only one case must match, which may require implementations to check all alternatives. Similarly, Property `&` (ampersand) expresses that a value matches several models simultaneously (**and**). This later operation is not useful when applied on tight objects as each model would be incompatible with the others, thus no object can match.

```
{ "season": { "|": [ "Spring", "Summer", "Fall", "Winter" ] },
  "movie": { "^": [ { "lang": "français", "titre":  "" },
                    { "lang": "Íslensk",  "titill": "" },
                    { "lang": "Runasimi", "suti":   "" } ] },
  "link": { "&": [ "$URI", "/^https:/" ] } }
```

Fig. 8. JSON Model with Combinations

Figure 8 illustrates logical combinators on an object with three properties. Property `season` is one of four strings. Property `movie` is a tagged union with Property `lang` as a discriminator, to provide a movie title. A JSON Model compiler can detect and take advantage of the discriminator to validate values without checking all cases (Sect. 4). Finally, Property `link` is an https URL.

To address the *convenient* use case of multiple inheritance which combines properties from different objects, we introduce the merge operator with Property `+` (plus). It must be understood as a pre-processing macro to merge directly-provided or referenced objects, which are then interpreted as a standard object. The merge operator is distributive over `|` and `^`, as used by Model `Elem` in Fig. 15. It cannot be distributive over `&` as it leads to contradictions. Property definitions are merged when they have the same name **and** the same model value. If one is mandatory and the other optional, the result is mandatory. Then optional reference and regex properties are merged, then the catch-all property. When model values are not equal, merging fails and the model is rejected.

```
{ "+": [
  { "!a": "", "?b": 0, "/^[a-z]+$/": "" },
  { "!b": 0, "?c": "", "": 0 } ] }
```

(a) Merging...

```
{ "!a": "", "!b": 0,
  "?c": "", "": 0,
  "/^[a-z]+$/": ""  }
```

(b) Merged

Fig. 9. JSON Model Merge Composition

Figure 9 Model (a) illustrates the merging of two objects definitions. In the resulting Model (b), each object contributes one mandatory property (`a` and `b`). First object optional Property `b` becomes mandatory because of the merge rule with `b` in the second object. Property `c` remains optional. Properties matching the regular expression must be strings (1st object). Other properties are also allowed if they are integers (2nd object).

```
{ "a": "Calvin",
  "b": 5432,
  "c": "R.03",
  "Age": 6        }
```

(a) Calvin

```
{ "a": "Susie",
  "b": 12345,
  "c": "R.02",
  "AGE": 7        }
```

(b) Susie

```
{ "a": "Hobbes",
  "B": 666,
  "c": "R.07",
  "age": 6        }
```

(c) Hobbes

Fig. 10. Three Sample Values

Figure 10(a) and (b) conform to the merged model, but (c) does not because mandatory `b` is missing (vs `B`) and `age` is lower case so should be a string.

3.6 Model Complements

This section addresses additional features: meta-data, defining and reusing elements, and importing external definitions.

Comments in a data structure description are *convenient* for understanding and maintenance. As JSON does not have a syntax for comments, we use Property # (sharp) for this purpose and for meta-data. Its value may be a string or an arbitrary object, allowing model designers to put any information they need. Property `version`, in the # section at the root of the model (Fig. 11), if set, denotes the JSON Model version of the model. Validator implementations may use properties in the object values for special configuration purposes, such as whether to consider `1.0` as an integer or not.

When developing large models, it is useful to factor out parts that can be reused consistently, avoiding repetitions (*composable*). As we already have a mechanism for referencing a model with a string ($), we only need to add new local definitions inside a model. Property % (percent), at the root of the model,

```
{ "#": { "name": "A Book", "version": 1 },
  "%": { "Section": { "title": "/./",
                      "?text": "",
                      "?sections": [ "$Section" ] } },
  "+": [ { "authors": [ "/./" ], "publisher": "/./" }, "$Section" ] }
```

Fig. 11. JSON Model for a Book

```
{ "title": "JSON Model: a Lightweight Featureful DSL for JSON",
  "authors": [ "Calvin", "Susie" ],
  "publisher": "Hobbes",
  "sections": [ { "title": "Introduction", "text": "JSON [10] ..." },
                { "title": "Related Work", "text": "As data st..." } ] }
```

Fig. 12. Example Book Data

introduces an object with properties as identifiers in a unique name space and values as the corresponding model. As a convention to enhance readability, predefined names are in uppercase, and user-defined names should be capitalized (*discernible*). For simple recursive structures, naming an element can also be done directly in its definition with Property $ (dollar) and the name as a value.

Figure 11 shows an *expressive* recursive data model, (`Section` uses itself as an array item in `sections`). The recursion is well-founded because Property `sections` is optional, or the list can be empty. Figure 12 shows a conforming value.

For large data structures, and to reuse already defined models, *modular*ity is achieve by referencing a model which is stored outside of the model definition. For this, URLs are allowed as a name in a reference ($), including a possible fragment, similarly to JSON Schema, e.g. `"$https://json-model.org/geom#Polygon"` references the `Polygon` model in the referenced model. Note that, unlike JSON Schema, it is intentionally not possible to use a path to reference an element in the model: only named elements can be reused.

```
{ "#": "Definitions at https://json-model.org/geom",
  "%": { "Coord": { "x": 0.0, "y": 0.0 },
         "Polygon": [ "$Coord" ]  } }
```

Fig. 13. Geometric Definitions

```
{ "%": { "Geo": "$https://json-model.org/geom" }, "pol": "$Geo#Polygon" }
```

Fig. 14. Geometric Use

Figures 13 and 14 show a definition used indirectly. Geometric types are defined in the `geom` model, and then imported and used in another model.

3.7 JSON Model Summary

JSON Model, introduced in [7,20], relies on 15 special keywords and 6 sentinel characters on top of JSON syntax, which is much lower than JSON Schema (60). It uses symbols with well-known semantics from programming languages, such as # for comments, | & + ^ for combinators, $ for references, /.../ for regex. Keywords are short to avoid mistyping. They use symbols to be language independent (*universal*) and easy to *discern* from the described data structure.

```
{
  "#": { "name": "Self-Validating JSON Model", "version": 1 },
  "%": {
    "Any": { "&": [ ] },
    "None": { "|": [ ] },
    "Val": { "|": [ null, true, -1, -1.0, "" ] },
    "Meta": { "|": [ "", { "": "$Any" } ] },
    "Array": [ "$Model" ],
    "Cons": { "_@": "$Model",
              "/^(<=|>=|<|>)$/": { "|": [ -1, -1.0, "" ] },
              "/^(=|!=)$/": "$Val", "?!": true },
    "Or": { "_|": "$Array" },
    "And": { "_&": "$Array" },
    "Xor": { "_^": "$Array" },
    "Merge": { "_+": "$Array" },
    "Obj": { "": "$Model", "/^[|@&^+]$/": "$None" },
    "Elem": { "+": [ { "?$": "", "?#": "$Meta" },
        { "|": [ "$Cons", "$Or", "$And", "$Xor", "$Merge", "$Obj" ] } ] },
    "Model": { "|": [ "$Val", "$Array", "$Elem" ] },
    "Root": { "+": [ { "?%": { "": "$Model" } }, "$Elem" ] }
  },
  "|": [ "$Val", "$Array", "$Root" ]
}
```

Fig. 15. JSON Model Self-Validating Tight Meta-Model

Figure 15 demonstrates that JSON Model is *self-hosted* with a self-validating tight meta-model. It astutely redefines predefs "$ANY" and "$NONE" not to use any predefs. The definition of `Obj` excludes special properties with a regex from appearing directly in objects, avoiding ambiguities. This meta model uses disjunctions, two merges and one conjunction, and is recursive on `Model`. It could be made tighter, e.g. by using regular expressions to restricts string constants, local definition identifiers or list predefs.

Seldom used features from JSON Schema [3,20] such as multiple-of, in-array containment and mime-types are discarded. They could be added with more constraint symbols, e.g. * ~ mime. Logical assertions are also ignored: This rarely used feature creates very hard to understand schemas and requires expensive evaluations for validation.

4 JSON Model Implementation

Proof-of-concept implementations of JSON Model are available [8] in Python and were used extensively in [20]. Compared to JSON Schema, JSON Model simple structure allows to generate definite functions for target elements, without having to care about the external context or keeping track of checked and unchecked elements. This locality feature allows the *efficient* compilation of models. Our implementation offers three validators: a direct model interpreter; a dynamic compiler which generates check functions on the fly (`lambda`); a static compiler which generates source code for these. It also includes partial model-to-schema and schema-to-model converters.

All validators share a common preprocessing step: Combinators are flattened to simplify the structure, possibly following definitions (inlining); Merge operators are distributed over **or** and **xor**-operators and eliminated by computing the resulting objects following property merging rules (Sect. 3.5); Partial evaluation simplifies empty or one-item combinators then propagates some values when appropriate; If possible, **xor** combinators are changed to less costly **or**.

```
{
  "%": {
    "A": { "p": 0 },
    "B": { "^": [ "$A", { "q": 0 }] },
    "C": { "+": [ { "r": 0 }, "$B" ] },
    "D": { "^": [ true, { "^": [ 1, 1.0 ] } ] }
  },
  "^": [ "$D",
       { "^": [ "$A", { "|": [ ] } ] } ]
}
```

(a) Before

```
{
  "%": {
    "A": { "p": 0 },
    "B": { "|": [ "$A", { "q": 0 } ] },
    "C": { "|": [ { "r": 0, "p": 0 },
                  { "r": 0, "q": 0 } ] },
    "D": { "|": [ true, 1, 1.0 ] }
  },
  "|": [ true, 1, 1.0, "$A" ]
}
```

(b) After

Fig. 16. JSON Model Preprocessing

Figure 16 illustrates these steps: `C` definition is expanded (`B` is inlined) to remove the merge combinator; `D` combinators are flatten; All **xor** combinators are switched to **or** because the combined models are structurally distinct; the root combinators are simplified thanks to partial evaluation and inlining, for instance `{ "|": [ ] }` is `"$NONE"` and thus can be removed from the **xor** list.

The compilers generate functions for each named types (`%`) and for the target root model, and handles recursion by re-selecting the function when needed. Several optimizations are implemented: Discriminant properties on **or** and **xor** combinators over objects are automatically detected, the generated code inlines the object definitions and takes advantage of the discriminant to call only one object check function; Duplicated models in **xor** combinators are checked once and eliminated directly; `{ "^": [ "$ANY", X ] }` is optimized as model X complement; Enums are detected to generate simple and efficient code. The generated code could be further optimized with partial evaluation. A particular attention is needed to generate accurate messages when a value is rejected, to identify the precise issue and where it occurs both in the value and the model.

```
{
  "#":
    "float/word",
  "^": [
    0.0,
    "/^[a-z]+$/i"
  ]
}
```

(a) Model

```python
import re2 as re
re_0 = re.compile("(?i)^[a-z]+$").search
def check_model(val: Any, path: str = "$") -> bool:
    res = isinstance(val, float) and val >= 0.0
    if not res:
        res = isinstance(val, str) and \
            re_0(val) is not None
    return res
```

(b) Python Generated Code

Fig. 17. JSON Model Static Compilation

Figure 17 illustrates a model and the code generated by our static compiler for one model: The regex is pre-compiled; The code relies on Python dynamic type introspection (`isinstance`) to handle any value.

5 JSON Model Integration with OpenAPI

Many web and mobile applications use the HTTP protocol to interact with a server, following REST architecture principles [13] to create, read, update, delete and search (CRUDS) data. OpenAPI [18] (formerly Swagger Spec) is a language-agnostic JSON or YAML descriptions of such interfaces, with JSON Schema [19] used to describe the type of exchanged JSON data. OpenAPI 3.1.0 (2021) extends the latest JSON Schema (2020–12) with some keywords, allows extension properties and defines new formats for usual computer integers and floats. Schema definitions appear as `/components/schemas` in the OpenAPI object, and schemas can be used directly in `MediaType` and `Parameter` as `schema` properties, to describe input (request) and output (response) data, either with a reference to a definition or as direct type descriptions, see Fig. 18(a).

For assessing qualitatively typical use cases, we have collected 21 API specifications ([21] in `corpus/OpenAPI`) from various sources, ranging from v2.0 to v3.1: 11 API focus on services, with a variety of HTTP methods, and 10 are dedicated to data access (get only), either public open data (6) or private specific data (4). It would be interesting to extend this corpus to study OpenAPI usage. JSON schemas were extracted, analysed [9] and converted automatically to JSON model with our tools [8]. These specs involve non-trivial nested objects and arrays, justifying the need for type specification and documentation, as suggested by frequent meta-data such as `description` and `examples`.

In this sample, format extensions (`int32`...) are frequent. Advanced features such as combinators and regex are occasional, and a few schemas include size constraints. Some schemas use many loose objects (unknown or mostly optional properties, null values), but in the context should really be tight. Confirming our earlier findings, the overall schema quality is low: Two files contain fully broken schemas (ebay, walmart), with similar but invalid types and structures, probably generated by the same tool; For 7 small schemas extracted from OpenAPI v3 own documentation, 3 have type errors (missing type or invalid format, reported in April 2024); On the 12 other real-world APIs, only 2 do not have any kind of type errors, and 2 others have minor errors (inconsistent type in examples). Altogether, two-thirds of schemas in our limited sample contain type errors.

```
components:
  schemas:
    Pet:
      type: object
      properties:
        id:
          type: integer
          format: int64
        name:
          type: string
      required:
      - id
      - name
      additionalProperties: false
paths:
  /pets/{petId}:
    get:
      parameters:
      - name: petId
        in: path
        required: true
        schema:
          type: integer
          format: int64
      responses:
        '200':
          content:
            application/json:
              schema:
                $ref: '#/components/schemas/Pet'
```

(a) with JSON Schema (30 lines)

```
components:
  models:
    Pet:
      id: $I64
      name: $STRING
paths:
  /pets/{petId}:
    get:
      parameters:
      - name: petId
        in: path
        required: true
        model: $I64
      responses:
        '200':
          content:
            application/json:
              model: $Pet
```

(b) with JSON Model (18 lines)

Fig. 18. YAML OpenAPI Comparison

OpenAPI is an important use case for describing JSON data structures. To integrate JSON Model *convenient*ly, we follow JSON Schema approach. As YAML is a superset of JSON, it can be used to represent models. Figure 18 compares OpenAPI with schemas and models. Models are declared in `components/models`, and specified in `model` properties. As expected, type definitions are much shorter (*compact*). For readability, we have used explicit type names instead of relying only on type inference. Moreover, for homogeneity, JSON Schema style references in OpenAPI could use a more direct JSON Model style reference.

6 Discussion and Future Works

JSON Model satisfies the 12 criteria about power, structure and user experience set for its design:

- It is nearly as *expressive* as JSON Schema, with only minor features missing, most of which could be integrated easily. External references allow to reuse models (*modular*, Fig. 14). Available compilers prove that validating models can be *efficient*. The overall power of JSON model is illustrated in Fig. 15 with a 22 nested objects *self-hosted* single-file meta-model, to be compared to 123 over 8 files for JSON Schema meta-schema.
- JSON Model examples are typically 4 times more *compact* than their JSON Schema or JSound counterparts (Figs. 2 & 3 vs 5 or Fig. 18(a) vs (b)). Even if the readability is debatable, thanks to type inference models remain very similar to the value they describe, which helps interpreting them. Models are *composable* to simplify model conception. *Frugal*ity is demonstrated by the use of only 19 keywords or sentinel characters compared to JSON Schema 60, which enables a rather small meta-model. However, predefs bring redundancy as a readability trade-off, e.g. compare `""` vs `"$STRING"`. The exclusive use of symbols makes it culturally independent (*universal*), possibly at the price of readability.
- *Intuitive*ness is judgemental in nature. Testing this would require careful experiments with potential users. However, as type inference is a guide for the language conception, models and values are similar, thus we think that it should be easier to understand than the large vocabulary of JSON Schema and its complex semantics. Many errors found in JSON Schema result from mixing schema description keywords (meta level) with actual schema keywords (user level). JSON Model solves this *discern*ibity issue by using symbols for descriptions, which makes it harder to confuse. People are very sensitive about the *tight* vs loose default behavior, which leads to endless and fruitless debates. An objective measure is to notice in Fig. 18 that tightening `Pet` schema definition requires 4 lines, whereas loosening a tight model would require only one. We believe that discernibility and tightness make models less error-prone, thus more secure. Finally, Fig. 5 shows the compact and *convenient* definition of a simple model, which is a good thing for OpenAPI integration.

Future work includes writing a formal specification, developing a JavaScript reference implementations, and distributing these widely with an illustrated tutorial and example use cases. Moreover, we would like to improve the qualitative evaluation of the language with user and case experiments, especially about intuitiveness, discernibility and convenience. More optimization could be implemented in the preprocessor and the compiler: **or** combinators could reorder models so that more frequent models are checked and matched earlier; model inclusions could be used to simplify some combinators. Performance evaluations of our implementations should be conducted.

Acknowledgements. We would like to thank Olivier Hermant for his help in proofreading this paper, Pierre Jouvelot for enlightening discussions, and the anonymous reviewers for their helpful comments.

References

1. Andrei, C., Florescu, D., Fourny, G., Robie, J., Velikhov, P.: JSound 2.0 – the complete reference (2018). https://jsound-spec.org
2. Attouche, L., Baazizi, M.A., Colazzo, D., Ghelli, G., Sartiani, C., Scherzinger, S.: Validation of modern JSON schema: formalization and complexity. In: Proceedings of POPL. ACM (2024)
3. Baazizi, M.-A., Colazzo, D., Ghelli, G., Sartiani, C., Scherzinger, S.: An empirical study on the "usage of not" in real-world JSON schema documents. In: Ghose, A., Horkoff, J., Silva Souza, V.E., Parsons, J., Evermann, J. (eds.) ER 2021. LNCS, vol. 13011, pp. 102–112. Springer, Cham (2021). https://doi.org/10.1007/978-3-030-89022-3_9
4. Baazizi, M.A., Colazzo, D., Ghelli, G., Sartiani, C., Scherzinger, S.: Negation-closure for JSON Schema. Theoret. Comput. Sci. **955**, 113823 (2023)
5. Bierman, G., Abadi, M., Torgersen, M.: Understanding typescript. In: Jones, R. (ed.) ECOOP 2014. LNCS, vol. 8586, pp. 257–281. Springer, Heidelberg (2014). https://doi.org/10.1007/978-3-662-44202-9_11
6. Boldt, J.: The common object request broker: architecture and specification. Specification Formal/97-02-25, Object Management Group (1995)
7. Coelho, F., Yannou-Medrala, C.: JSON model: a lightweight featureful description language for JSON data structures. Technical report A/795/CRI, CRI, Mines Paris – PSL (2023)
8. Coelho, F., Yannou-Medrala, C.: JSON model implementation (2023–2024). https://github.com/clairey-zx81/json-model
9. Coelho, F., Yannou-Medrala, C.: JSON schema statistics tools (2023–2024). https://github.com/clairey-zx81/json-schema-stats
10. Crockford, D.: The application/JSON media type for javascript object notation (JSON). RFC 4627, IETF (2006)
11. Damas, L., Milner, R.: Principal type-schemes for functional programs. In: PoPL, pp. 207–212 (1982)
12. ECMA International: Standard ECMA-262 – ECMAScript Language Specification. 5.1 edn., 846 p. (2011). First edition in 1999
13. Fielding, R.T.: REST: architectural styles and the design of network-based software architectures. Doctoral dissertation, University of California, Irvine (2000)
14. Gruber, J., Swartz, A.: Markdown. Specification 1 (2004)
15. Ingerson, B., Evans, C., Ben-Kiki, O.: Yet another markup language (YAML). Specification 1.0 (2001)
16. Jouvelot, P., Triolet, R.: NewGen: a language-independent program generator. EMP-CRI 191, CRI, Mines Paris – PSL (1989)
17. Mitre: Inefficient regular expression complexity. CWE 1333 (2021)
18. OAS: OpenAPI. Specification v3.1.0, Linux Foundation (2021). https://spec.openapis.org/
19. Wright, A., Andrews, H., Hutton, B., Dennis, G.: JSON schema: a media type for describing JSON documents. Draft 2020-12, IETF (2022)
20. Yannou-Medrala, C., Coelho, F.: An analysis of defects in public JSON schemas. In: BDA 2023: 39ème Conference sur la gestion de données – Principes, technologies et applications (2023)
21. Yannou-Medrala, C., Coelho, F.: Yet another JSON schema corpus (2023–2024). https://github.com/clairey-zx81/yac

A Data Model and Predicate Logic for Trajectory Data

Johann Bornholdt(✉), Theodoros Chondrogiannis, and Michael Grossniklaus

University of Konstanz, 78457 Konstanz, Germany
{johann.bornholdt,theodoros.chondrogiannis, michael.grossniklaus}@uni-konstanz.de

Abstract. With recent sensor and tracking technology advances, the volume of available trajectory data is steadily increasing. Consequently, managing and analyzing trajectory data has seen significant interest from the research community. The challenges presented by trajectory data arise from their spatio-temporal nature as well as the uncertainty regarding locations between sampled points. In this paper, we present a data model that treats trajectories as first-class citizens, thus fully capturing their spatio-temporal properties. We also introduce a predicate logic that enable query processing under different uncertainty assumptions. Finally, we show that our predicate logic is expressive enough to capture all spatial and temporal relations put forward by previous work.

Keywords: Trajectory Data · Data Modeling · Predicate Logic

1 Introduction

A growing number of applications ranging from rating and publishing personal hiking trips [7] to studying the migration of animals require the analysis of trajectory data. Consequently, the efficient processing of trajectory data has attracted significant interest [6,26]. For example, at the Centre for the Advanced Study of Collective Behaviour[1] at the University of Konstanz, the excellence cluster in which the presented research is situated, we are building the so-called *Imaging Hangar*, which enables us to study small animal collectives in a controlled environment using trajectory data obtained from video image analysis [25]. Depending on the kind of object being tracked, the data recorded together with a trajectory is highly application-specific. Furthermore, the quality of the trajectory data can vary substantially based on sampling rate and sensor accuracy.

Trajectory data are uncertain by nature. Specifically, at least two types of uncertainty can be distinguished. The first type comes from noise in GPS measurements and is inherent to the data source, making it impractical to address it at the system level. The second type concerns the position of an object between

[1] https://www.exc.uni-konstanz.de/collective-behaviour/.

J. Tekli et al. (Eds.): ADBIS 2024, LNCS 14918, pp. 18–31, 2024.
https://doi.org/10.1007/978-3-031-70626-4_2

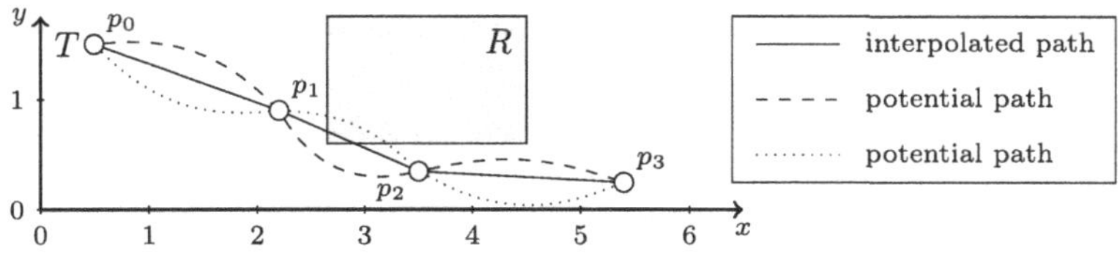

Fig. 1. Example of a trajectory T and a query region R.

two consecutive trajectory points and is the focus of the work presented in this paper. Due to the discrete sampling rate with which locations are obtained, there is uncertainty as to the exact movement of an object at every point in time. For example, Fig. 1 shows the trajectory of a bird T and a query region R. Given a straight line between points p_1 and p_2, T intersects R. However, the bird could have actually moved around the corner of R, shown as a dashed line.

Existing systems like SECONDO [15] and MobilityDB [31] come with two major shortcomings. First, they assume that all necessary trajectories can be collected and stored in a single location. Such a case is not always possible as exchanging trajectory data obtained from different sources has both practical and legal limitations. Second, to deal with uncertainty, existing approaches either model trajectories using cylinders [24] and beads [21], or attempt to process queries by inferring the exact location of the moving object between two recorded locations [28, 29]. However, existing systems do not take the uncertain nature of trajectory data into account.

To address these shortcomings, our aim is to develop a query broker that enables users such as biologists and environmental scientists to query trajectory data from multiple sources through a unified interface. As a first step, this paper proposes a spatio-temporal predicate logic, including a data model and operators, to query trajectories from different sources. In particular, our predicate logic that accommodates uncertainty in trajectory data by supporting different levels of strictness. The contributions of this paper are as follows.

- We introduce a data model for trajectories based on the NF^2 relational data model. Our model gives equal importance to the spatial and temporal attributes of trajectories while also supporting their application-specific attributes (Sect. 3).
- We define a unified spatio-temporal predicate logic to express selection operations over trajectory data. As a distinguishing feature, our predicate logic supports different levels of strictness to deal with uncertainty in interpreting trajectory data (Sect. 4).
- We demonstrate that our spatio-temporal predicate logic is expressive enough to represent the spatial relations from the DE-9IM standard [11] and the temporal relations from Allen's Interval Algebra [1], and we show how our logic handles the uncertainty of trajectory data in a query (Sect. 5).

Section 2 provides an overview of existing works. Concluding remarks and directions for future work are given in Sect. 6.

2 Related Work

In this section, we provide an overview of existing data models, algebras, and systems that have been proposed to store and query trajectory data.

Data Models and Algebras. There has been a variety of contributions in the field of data models and algebras for trajectories and moving objects [3,4,10, 18,22]. Güting et al. [16] provide a foundational framework for representing and querying moving objects, which serves as a cornerstone in the trajectory data management domain. Frihida et al. [13] introduce an algebraic spatio-semporal trajectory data type for the representation of trajectory data. Building on the approach of Frihida et al., Zheni et al. [30] introduce a semantic-based model and manipulation language for trajectories. A contribution by Ferreira et al. [12] presents an algebra for trajectories by incorporating time series and coverage. Bakli et al. [2] propose an algebra on operators based on the Hadoop system. In contrast to our work, these contributions do not deal with the uncertainty between the sampled points of trajectories in the data model or the algebra.

Systems. Several works have contributed to the field of moving object data management [8,9,19,26,27]. Notably, DEDALE [14] is an early system that laid the groundwork for representing and querying moving objects. DEDALE serves as a spatial extension for SQL, lacking a temporal component. SECONDO [15] is a research prototype that implements a subset of the foundational framework proposed by Güting et al. [16]. The HERMES trajectory database engine [20] extends the object-relational data model by introducing data types and DDL extensions for managing trajectory data. MobilityDB [31] is an extension to PostGIS that supports moving object data providing trajectory-specific data types and functions that implement the DE-9IM relations to a certain degree. UlTraMan [9] extends Apache Spark offering a holistic solution for the entire trajectory pipeline, including range query processing. Moreover, while most of the aforementioned systems focus primarily on the spatial dimension of trajectory data, time-series database systems [17] also support storing and querying trajectory data, focusing primarily on the temporal dimension. In contrast to our work, most of these systems do not come with a formally defined data model, but instead focus on the technical challenges of trajectory data management. Furthermore, existing systems do not take into account the uncertain aspects of trajectory data.

3 A Data Model for Trajectories

Trajectories represent the movement of a moving object. Typically, trajectories are given as a sequence of tuples, that consist of geometric or geographical coordinates accompanied by timestamps. The timestamp attributes of the tuples of trajectories are strictly ordered and monotonically increasing. Since most currently available datasets are two-dimensional, we focus on two spatial dimensions. Definition 1 defines a trajectory in a relational context.

order	x	y	τ
0	1.0	0.5	110
1	2.0	1.0	120
2	4.0	1.5	130
3	4.0	1.5	140
4	3.0	0.5	150

Fig. 2. The relational representation of trajectory T.

tid	T			
	order	x	y	τ
T_0	0	1.0	0.5	110
	1	2.0	1.0	120
	2	4.0	1.5	130
	3	4.0	1.5	140
	4	3.0	0.5	150

Fig. 3. Relational representation $\mathfrak{T}$ of a nested trajectory relation T_0.

Definition 1. *Let trajectory relation T be a relation with schema $sch(T) = (o, x, y, \tau)$ that satisfies the following:*

- $val(T) = \{(o_i, x_i, y_i, \tau_i) \mid i \in [0, n] \wedge n \in \mathbb{N}\}$,
- *o is the order of each tuple $tp \in T$.*
- *x and y are spatial coordinates (geometrical or geographical).*
- *τ are timestamps.*
- *$\forall tp_i, tp_j \in T, i \neq j$ it stands that $tp_i(o) > tp_j(o) \Leftrightarrow tp_i(\tau) > tp_j(\tau)$.*

In the trajectory relation, we include the *order* column. While the timestamp could also be used to determine the correct sequence of the tuples in the relation, the order facilitates specific operations, e.g., the retrieval of line segments between consecutive points. Figure 2 shows T in a relational table. It is helpful to store trajectories in relational tables because RDBMS offer a multitude of operators that can be used to run queries on the trajectory data.

3.1 Trajectory Representation in NF^2

The *Non-First Normal Form* (NF^2) data model is an extension of the relational data model. The corresponding NF^2 algebra [23] allows subexpressions as predicates and enables the access of nested relations. In the context of our work, the NF^2 data model enables the modeling of trajectories as nested relations, thus treating them as first-class citizens. More specifically, using the NF^2 data model we store single trajectories T as nested relations of a *trajectories relation* $\mathfrak{T}$, i.e.,

$$\mathfrak{T}(\underline{tid}, T(\underline{order}, x, y, \tau))$$

Figure 3 shows an example of a trajectories relation with a single nested relation.

3.2 Data Point and Trajectory Properties

In order to represent additional properties of a trajectory, we can add a relation with properties referencing the trajectory relation. For properties that apply to entire trajectories, we add a column to the properties relation for each property. Trajectory properties can be added with two different scopes:

1. *trajectory properties*, for properties on entire trajectories.
2. *point properties*, for properties on trajectory points.

begin	end	movement type
0	1	walking
2	4	flying

Fig. 4. Property relation for trajectories (species) and points (movement type).

begin	end	movement type
0	1	walking
2	4	flying

Fig. 5. Query result when interpreting consecutive point properties continuously.

Figure 4 shows a trajectory property relation with one example column for each property type. A *trajectory property* is shown in the *species* column. It contains the type of animal, that was tracked for this trajectory. As a trajectory property, its value applies to the entire trajectory. A *point property* enables the storage of a specific property associated with a single point of the trajectory. The column storing point properties contains a nested relation with the *order* of the point and the corresponding property. It is important for the consistency of the data model, to relate it to the order instead of the timestamp, to enforce that properties always correspond to specific points of the trajectory. Point properties can be helpful for properties that only apply to a few points. For example, in the properties relation shown in Fig. 4, the *movement type* column contains information about how the bird is moving, e.g., flying or walking.

3.3 Segment Property Uncertainty

An inherent problem when representing the movement of an object using trajectories is that observations can only be captured at distinct timestamps. However, the actual movement of an object is continuous. As such, when processing a query, we cannot assume that any point properties also apply to locations on the segments between two consecutive points. Figure 5 shows a segment interpretation of the point properties in Fig. 4. Between points 0 and 1, i.e., between timestamps 110 and 120, the movement type property has the value "walking" and between points 2 and 4, i.e., timestamps 130 and 150, the value "flying". Hence, whether the value "walking", applies for timestamp 115 depends on the semantics of the point property. Even so, no assumption can be made about the value of the property, e.g., for the timestamp 125, i.e., it is uncertain at which point in time the value of the property changes. Due to this uncertainty, we do not consider this type of property in the data model. Instead, we consider such properties only in the context of a query and we propose a strictness option in Sect. 4.4 to deal with the uncertainty.

4 Spatio-Temporal Predicate Logic

The selection operator is an essential query operator in database systems that filters tuples based on a given predicate. In this section, we introduce the spatio-temporal selection σ^{ST} operator that performs a range selection over a set of trajectories on the spatial and temporal dimensions. The operator applies a spatio-temporal predicate P on a relation of trajectories $\mathfrak{T}$. The result is a subset of the trajectories in $\mathfrak{T}$ which satisfy P.

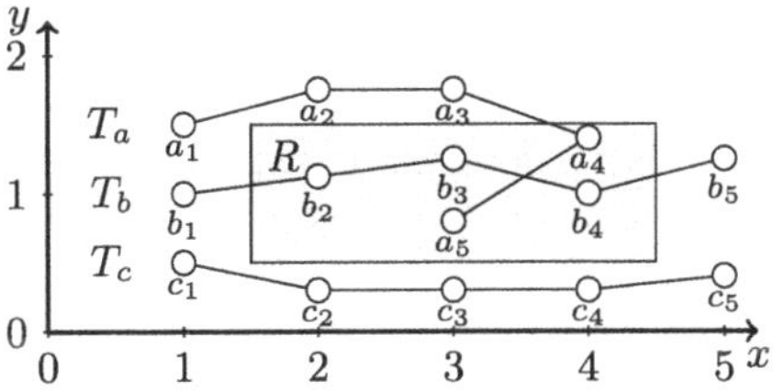

Fig. 6. Spatial component of Q

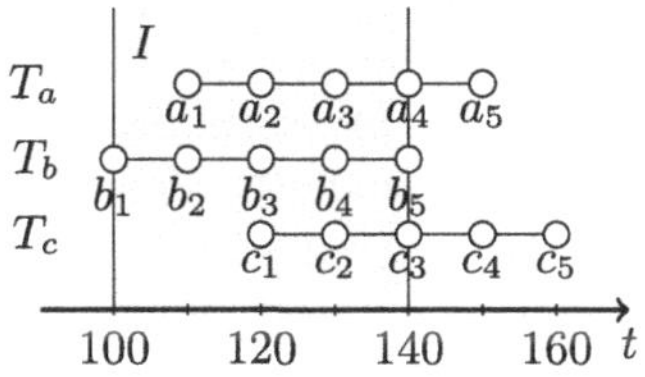

Fig. 7. Temporal component of Q

Definition 2. *Given a trajectories relation $\mathfrak{T}$, and a spatio-temporal predicate P, the spatio-temporal trajectory selection σ^{ST} returns a relation that contains every tuple tp in $\mathfrak{T}$ for which the trajectory relation $tp.T$ satisfies P, i.e.,*

$$sch(\sigma_P^{ST}(\mathfrak{T})) = sch(\mathfrak{T})$$
$$val(\sigma_P^{ST}(\mathfrak{T})) = \{tp : tp \in \mathfrak{T} \wedge P \text{ holds for } tp.T\}$$

To express spatio-temporal predicates, we have designed a predicate notation that works equally for the spatial and temporal dimensions. Our notation can be used to define specific conditions on the points of trajectories. When using a predicate in a selection operator, all of these conditions must be satisfied for a trajectory to be in the result set. Additionally, multiple predicates can be combined in conjunctive normal form. For example, let a query Q be: "find all trajectories in an region R during a time interval I". Figures 6 and 7 show a visualization of the spatial and temporal portion of the query, respectively. In the following subsections, we introduce our predicate logic and show how it can independently solve the spatial and temporal parts of Q.

4.1 Spatial Predicates

The spatial part of a predicate is used to express a two-dimensional range query. The predicates are applied to trajectories on a point level. The following objects can be used in spatial predicates:

- T: The trajectory relation that contains all points in the trajectory,
- R: A geometric object,
- $\{p_f, p_l\} \in T$: The first and last point of the trajectory,
- $T_{\widehat{fl}} = T \setminus \{p_f, p_l\}$: All points of the trajectory except the first and the last.

Without loss of generality, we consider the case of spatial range queries, where R represents a region. While on a logical level, R can be any two-dimensional shape, the nature of the physical implementation of the predicates can affect the supported shapes of R. Furthermore, the calculation of the predicates changes depending on whether R is, for example, a rectangle or a polygon. Since the most common spatial range queries are bounding boxes, we focus on rectangle-shaped regions in the upcoming examples.

The relationship between points of a trajectory T and R can be expressed with the following operators: *contained* ($\sqsubseteq$), *properly contained* ($\sqsubset$), and *not contained* ($\not\sqsubseteq$). The *contained* operator asserts that one or several points must be inside R or on the border of R. It can be used to model the spatial portion of query Q shown in Fig. 6 by asserting that at least one point in T must be contained in R with the predicate:

$$\exists p \in T : p \sqsubseteq R$$

In order to ensure that the entire trajectory T is contained in (or on the border of) R, a predicate can be written as:

$$\forall p \in T : p \sqsubseteq R$$

The *properly contained* operator functions similarly to *contained* but excludes points that lie on the border of R. In addition to expressing that points are properly contained in R, *properly contained* can also be used in conjunction with *contained* to express that points are on the border of R:

$$\forall p \in T : p \sqsubseteq R \wedge \neg(p \sqsubset R)$$

Lastly, the *not contained* operator enables the construction of predicates for points that are disjoint from R:

$$\forall p \in T : p \not\sqsubseteq R$$

For some complex spatial relations, it is necessary to address a trajectory's first or last parts specifically. To achieve this, we introduce two unique points, p_f and p_l, denoting the trajectory's first and last points T, respectively. These special points can be used, for example, to express that the beginning of T is contained in R, but the end of T is not:

$$p_f \sqsubset R \wedge p_l \not\sqsubseteq R$$

To enhance the conciseness of the formulas, we utilize the subset $T_{\widehat{fl}}$ of T, which includes all points except p_f and p_l. Using $T_{\widehat{fl}}$, a predicate for a trajectory that starts and ends inside R, but has points outside of R is expressed as:

$$p_f, p_l \sqsubset R \wedge \exists p \in T_{\widehat{fl}} : p \not\sqsubseteq R$$

4.2 Temporal Predicates

The operators introduced in Sect. 4.1 ($\sqsubset$, $\sqsubseteq$, $\not\sqsubseteq$) for spatial predicates can also be used to express the relationship between trajectories points and a time interval I.

In contrast to the two-dimensional region R, I is one-dimensional, which makes the start and endpoints of I its *border*. Figure 7 shows the temporal part of query Q, where the border points of I are at timestamps 1 and 4. The temporal part of Q can be expressed with the introduced operators:

$$\exists p \in T : p \sqsubseteq I$$

To effectively express temporal relations, we introduce two additional operators: p *is before* I $(p < I)$, and p *is after* I $(p > I)$. The operator $p < I$ expresses that point p is earlier on the time axis than the start of the interval I, for $p > I$, p is after the end point of I. Therefore, both operators express that p is outside of I. For example, a predicate for a trajectory that completely overlaps interval I can be expressed as:

$$p_f < I \wedge p_l > I$$

4.3 Combining Spatial and Temporal Predicates

One strength of our predicate logic is its ability to combine spatial and temporal predicates seamlessly. Users can express complex spatio-temporal queries involving the geometric characteristics of trajectories and their temporal evolution.

When looking at the example query q, we can now express the spatial and temporal parts of the query. However, suppose we express both independently from each other. In that case, we can see in Figs. 6 and 7 that T_a intersects both spatial region R, and interval I, but not in the same points. To properly express Q, we need to define a predicate where a single point p of trajectory T is both in R and I:

$$\exists p \in T : p \sqsubseteq R \wedge p \sqsubseteq I$$

In our predicate logic, spatio-temporal predicates can be expressed easily because the same operators can be used on both the spatial and the temporal dimensions.

4.4 Selection Uncertainty

As discussed in Sect. 3.3, the segments between individual points are not known in trajectories. For the spatio-temporal selection, these unknown segments pose a problem because the predicate operators introduced in Sects. 4.1 and 4.2 are applied on sets of points. However, when using the concrete points of trajectories, we are not examining the segment between points.

Figure 1 shows a bird's trajectory T and a query region R. To check whether a trajectory T intersects with a region R, assume a predicate $Q_1 = \exists p \in T : p \sqsubset R$, which checks if T has a point which is contained in R. On a point-by-point evaluation, T does not satisfy q_1 because all points of T lie outside of R, even though we can see in Fig. 1 that the interpolation of T intersects R. The same uncertainty also exists in the inverse query, when checking whether T does not intersect R with the predicate $Q_2 = \forall p \in T : p \not\sqsubseteq R$.

To tackle this uncertainty in the spatio-temporal predicates, we propose a *strictness parameter*, added to the predicates, to define how ambiguous trajectories will be treated for each predicate. We identify three degrees of strictness:

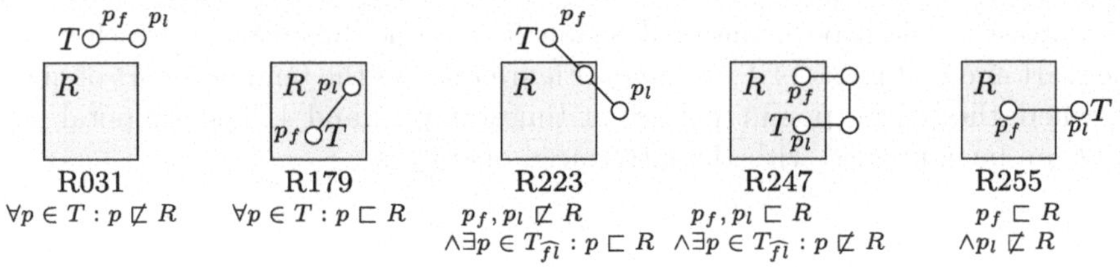

Fig. 8. Visualizations of DE-9IM relationships with their equivalent predicates.

- The *strict* evaluation of predicates considers only the points of the trajectory. In the example above, the strict evaluation of Q_1 does not match T, while Q_2 does match T.
- The *relaxed* evaluation of predicates assumes that there are infinitely many intermediate points on the straight line segments between pairs of consecutive points in trajectories, thereby assuring that all intersections between T and R are considered. In the example above, the relaxed evaluation of Q_1 matches T, while Q_2 does not match T.
- The *approximated* evaluation allows users to inject custom behavior into the predicates for cases where the *strict* and *relaxed* evaluations are not sufficient, e.g., for restricted trajectories.

With the strictness parameter, the user can define in a query whether the trajectory should be evaluated as a set of points (strict), as a continuous movement (relaxed), or by using some user-define assumption about the movement of the object (approximated). As investigating multiple assumptions for the *approximated* evaluation is out of the scope of our paper, we focus on strict and relaxed evaluation. This strictness parameter applies for spatial predicates as well as for the segment properties described in Sect. 3.3.

5 Proof of Concept

In this section, we demonstrate the completeness of our predicate operators, introduced in Sect. 4 for spatio-temporal range queries by establishing their equivalence and compatibility with established relationship models.

Regarding spatial relations, we consider the *Dimensionally Extended Nine-Intersection Model* (DE-9IM) [11], a widely utilized topological model to define and reason about spatial relationships between geometric shapes. The model defines nine intersection patterns regarding the *interior*, *boundary*, and *exterior* between two geometric objects in two dimensions to characterize their spatial relation. Note that in case one of the geometric objects as a linestring, e.g., the spatial component of a trajectory, the interior is the linestring itself, and the boundary is empty. While the DE-9IM consists of 6,561 distinct relations between pairs of shapes, Zlatanova et al. [32] demonstrate that only 19 relationship types are necessary to model all possible relationships between polygons and linestrings. Since in our problem setting spatial range queries are relationships

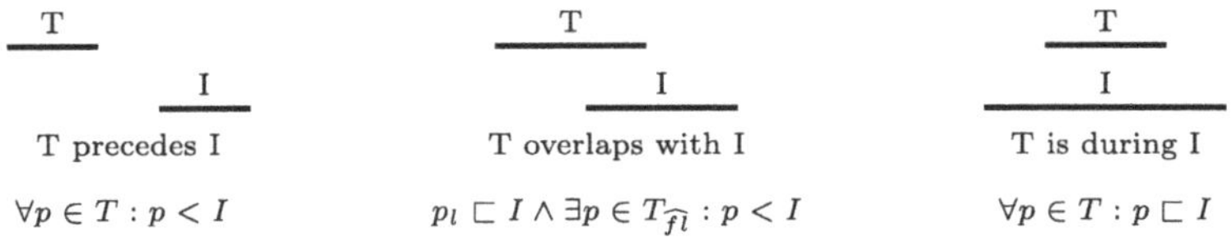

Fig. 9. Example visualizations of time intervals with their equivalent predicates.

between polygons and linestrings, our predicate logic can be considered complete because it can express the 19 relations mentioned above. Figure 8 demonstrates, with five examples, how DE-9IM relationships can be expressed as predicates.

Regarding temporal relations, we focus on *Allen's Interval Algebra* [1], which provides a formal framework for representing and reasoning about temporal intervals. The algebra defines thirteen possible binary relations between time intervals. With our temporal predicates it is possible to express all interval relations defined in Allen's Interval algebra. Figure 9 shows three examples of interval relations expressed as predicates. Due to space limitations, for the complete list of DE-9IM and Allen's Interval Algebra relations along with the corresponding predicates, we refer to the extended version of our paper [5].

As a proof of concept, we consider the spatio-temporal range query and we show how we can express the spatial and temporal predicates in the NF^2 algebra.

Definition 3 (Spatio-temporal Range Query). *Given a trajectories relation $\mathfrak{T}(\underline{tid}, T(\underline{order}, x, y, \tau))$ a spatial region R, a time interval I, a spatial predicate P_s, and a temporal predicate P_τ, a spatio-temporal range query returns all tuples $tp \in \mathfrak{T}$ such that the sequence of x-y coordinates of $tp.T$ and R satisfy P_s, and the sequence of timestamps τ of $tp.T$ and I satisfy P_τ.*

For instance, the relation containing the set of x-y coordinates of a trajectory with $tid = 1$ is

$$\pi[x, y](\mu_T(\sigma[tid{=}1](\pi[tid, \pi[x, y](T)](\mathfrak{T})))).$$

Spatial Trajectory Selection in NF^2. We begin by showing how NF^2 algebra can be used to answer queries involving spatial predicates. Note that relationships which involve points or lines lying on the border of a query region are not very useful in practice. Hence, due to the limited space, we focus on the five relationships that do not involve points or lines on the border of the query region, i.e., R031, R179, R223, R247, and R255[2]. Without loss of generality, we consider the query region to be a rectangle $R = (x_{min}, y_{min}, x_{max}, y_{max})$.

We begin with relations R179, R247, and R255. For these relations, the algebraic expressions for both strict and relaxed evaluation are the same. Wherever necessary in the following examples we have:

$$\begin{aligned} P_{first} &= \sigma[order = 1](T) \\ P_{last} &= \sigma[order = \max(\pi[order](T'))](T) \end{aligned}$$

[2] The DE-9IM relation numbers are based on Zlatanova et al. [32].

Example 1. Given a trajectories relation $\mathfrak{T}$ and a query rectangle R, the relationship R179 returns all trajectories/tuples $tp \in \mathfrak{T}$ such that the trajectory T lies completely inside R. The NF^2 algebra expression for this query is

$$\sigma^{ST}_{R179,R} = \sigma[x_{min} < \min(\pi[x](T)) \wedge y_{min} < \min(\pi[y](T)) \wedge$$
$$x_{max} > \max(\pi[x](T)) \wedge y_{max} > \max(\pi[y](T))\ ](\mathfrak{T})$$

Example 2. Given a trajectories relation $\mathfrak{T}$ and a query rectangle R, the relationship R247 returns all trajectories $tp \in \mathfrak{T}$ the starting point p_f and the ending point p_l lie completely inside R and there exists at least one point p of t that lies outside R. The NF^2 algebra expression for this query is

$$\begin{aligned}\sigma^{ST}_{R247,R} = \sigma[&\pi[x](P_{first}) > x_{min} \wedge \pi[x](P_{first}) < x_{max} \wedge\\ &\pi[y](P_{first}) > y_{min} \wedge \pi[y](P_{first}) < y_{max} \wedge\\ &\pi[x](P_{last}) > x_{min} \wedge \pi[x](P_{last}) < x_{max} \wedge\\ &\pi[y](P_{last}) > y_{min} \wedge \pi[y](P_{last}) < y_{max} \wedge (\\ &\quad \min(\pi[x](T)) < x_{min} \vee \max(\pi[x](T)) > x_{max} \vee\\ &\quad \min(\pi[y](T)) < y_{min} \vee \max(\pi[y](T)) > y_{max})\ ](\mathfrak{T})\end{aligned}$$

Example 3. Given a trajectories relation $\mathfrak{T}$ and a query rectangle R, the relationship R255 returns all trajectories $tp \in \mathfrak{T}$ the starting point p_f of which lies inside R and the ending point p_l of which lies outside R. The NF^2 algebra expression for this query is

$$\begin{aligned}\sigma^{ST}_{R255,R} = \sigma[&\pi[x](P_{first}) > x_{min} \wedge \pi[x](P_{first}) < x_{max} \wedge\\ &\pi[y](P_{first}) > y_{min} \wedge \pi[y](P_{first}) < y_{max} \wedge (\\ &\quad \min(\pi[x](P_{last})) < x_{min} \vee \max(\pi[x](P_{last})) > x_{max} \vee\\ &\quad \min(\pi[y](P_{last})) < y_{min} \vee \max(\pi[y](P_{last})) > y_{max})\ ](\mathfrak{T})\end{aligned}$$

We now focus on relations R031, and R223. As shown in Fig. 8, in order to apply the relaxed evaluation of these predicates, one must check every line segment form by consecutive trajectory points. In order to enable the examination of trajectory line segments, we define the following relation that we use in the subsequence examples:

$$T_{sgmt} = \pi[T.order, T.x, T.y, T'.x, T'.y](T \bowtie_{T.order+1=T'.order} T')$$

Example 4. Given a trajectories relation $\mathfrak{T}$ and a query rectangle R, the relationship R031 returns all trajectories $tp \in \mathfrak{T}$ that lie completely outside R. For the strict evaluation, it is sufficient to check that all points of each T lie outside R. As such, the NF^2 algebra expression for this query is

$$\sigma^{ST}_{R031,R} = \sigma[\mathrm{count}(\sigma[x_{min} < x \wedge y_{min} < y \wedge x_{max} > x \wedge y_{max} > y](T)) = 0\ ](\mathfrak{T})$$

For the relaxed evaluation, we also need to check that none of the segments of T in T_{sgmt} intersect R.

Example 5. Given a trajectories relation $\mathfrak{T}$ and a query rectangle R, the relationship R223 returns all trajectories $tp \in \mathfrak{T}$ the starting point p_f and the ending point p_l of the associated trajectory T that lie outside R and T intersects R. For the strict evaluation, it is sufficient to check that p_f and p_l lie outside R and there is at least one point of T that lies inside R. As such, the NF^2 algebra expression for this query is

$$\begin{aligned}\sigma^{ST}_{R223,R} =\sigma[(\ &\pi[x](P_{first})<x_{min} \vee \pi[y](P_{first})<y_{min}\vee \\ &\pi[x](P_{first})>x_{max} \vee \pi[y](P_{first})>y_{max})\wedge \\ &(\pi[x](P_{last})<x_{min} \vee \pi[y](P_{last})<y_{min}\vee \\ &\pi[x](P_{last})>x_{max} \vee \pi[y](P_{last})>y_{max})\wedge \\ &\mathrm{count}(\sigma[x>x_{min} \wedge y>y_{min} \wedge x<x_{max} \wedge y<y_{max}](T))>0\ ](\mathfrak{T})\end{aligned}$$

For the relaxed evaluation, in the case where count $= 0$ we need to check whether at least one of the segments in T_{sgmt} intersects R.

Temporal Trajectory Selection in NF^2. We now show how the NF^2 algebra can be used to answer queries involving temporal predicates. Due to the limited space, similar to the spatial trajectory selection, we focus on relationships that do not consider points lying at the start or the end of a given interval. Hence, we focus on the relationships *precedes*, *overlaps with*, and *is during*. Without loss of generality, we consider the query interval $I = (\tau_s, \tau_e)$.

Example 6. Given a trajectories relation $\mathfrak{T}$ and a query interval I, the *precedes* relationship returns all trajectories $tp \in \mathfrak{T}$ with a last point that has a timestamp before the starting timestamp τ_s of the interval I. The NF^2 algebra expression for this query is

$$\sigma^{ST}_{precedes,I} = \sigma[\max(\pi[\tau](T)) < \tau_s](\mathfrak{T}).$$

Example 7. Given a trajectories relation $\mathfrak{T}$ and a query interval I, the *overlaps with* relationship returns all trajectories $tp \in \mathfrak{T}$ for which the timestamp of the first point is before the start τ_s of the interval I, and the timestamp of the last point is after τ_s but before the end τ_e of the interval I. The NF^2 algebra expression for this query is

$$\sigma^{ST}_{overlaps\ with,I} = \sigma[\min(\pi[\tau](T)) < \tau_s \wedge \max(\pi[\tau](T)) > \tau_s \wedge \min(\pi[\tau](T))<\tau_e](\mathfrak{T}).$$

Example 8. Given a trajectories relation $\mathfrak{T}$ and a query interval I, the *is during* relationship returns all trajectories $tp \in \mathfrak{T}$ for which both the timestamp of the first point and the timestamp of the last point are after the start τ_s of the interval I and before the end τ_e of I. The NF^2 algebra expression for this query is

$$\sigma^{ST}_{overlaps\ with,I} = \sigma[\min(\pi[\tau](T)) > \tau_s \wedge \max(\pi[\tau](T)) < \tau_e](\mathfrak{T}).$$

In a similar fashion, we can define the mirrored relationships *is preceded by*, *is overlapped by*, and *contains*. Note that for the temporal dimension, there is no difference between the strict and the relaxed evaluation since we only need to consider the start and the end timestamp of a trajectory.

6 Conclusion

This paper proposes a formal data model and predicate logic for unified trajectory data management. Introducing a novel data model rooted in the NF^2 relational data model, we merge spatial, temporal, and application-specific attributes of trajectories. Our unified spatio-temporal predicate logic handles uncertainty in sampled trajectory data by accommodating varying levels of strictness.

Regarding future work, this paper lays the formal foundation for a query broker for trajectory data. Additionally, we aim to design a specialized query optimization framework to improve the broker's efficiency and scalability in handling complex trajectory queries, paving the way for enhanced trajectory data management and analysis capabilities.

Acknowledgment. This work is funded by the Deutsche Forschungsgemeinschaft (DFG) under Germany's Excellence Strategy – EXC 2117 – 422037984.

References

1. Allen, J.F.: Maintaining knowledge about temporal intervals. Commun. ACM **26**(11), 832–843 (1983)
2. Bakli, M.S., Sakr, M.A., Soliman, T.H.A.: A spatiotemporal algebra in Hadoop for moving objects. Geo-Spat. Inf. Sci. **21**(2), 102–114 (2018)
3. Bębel, B., Morzy, M., Morzy, T., Królikowski, Z., Wrembel, R.: OLAP-like analysis of time point-based sequential data. In: Castano, S., Vassiliadis, P., Lakshmanan, L.V., Lee, M.L. (eds.) ER 2012. LNCS, vol. 7518, pp. 153–161. Springer, Heidelberg (2012). https://doi.org/10.1007/978-3-642-33999-8_19
4. Bogorny, V., Kuijpers, B., Alvares, L.O.: A spatio-temporal data mining query language for moving object trajectories. UFRGS, TR-357 (2008)
5. Bornholdt, J., Chondrogiannis, T., Grossniklaus, M.: A data model and predicate logic for trajectory data (extended version). arXiv.org (2024)
6. Chondrogiannis, T., Bornholdt, J., Bouros, P., Grossniklaus, M.: History oblivious route recovery on road networks. In: ACM SIGSPATIAL, pp. 1–10 (2022)
7. Chondrogiannis, T., Ge, M.: Inferring ratings for custom trips from rich GPS traces. In: LocalRec@ACM SIGSPATIAL, pp. 1–4 (2019)
8. Cudre-Mauroux, P., Wu, E., Madden, S.: Trajstore: an adaptive storage system for very large trajectory data sets. In: IEEE ICDE, pp. 109–120 (2010)
9. Ding, X., Chen, L., Gao, Y., Jensen, C.S., Bao, H.: Ultraman: a unified platform for big trajectory data management and analytics. Proc. VLDB Endow. **11**(7), 787–799 (2018)
10. Düntgen, C., Behr, T., Güting, R.H.: Assessing representations for moving object histories. FernUniv., Fachbereich Informatik (2010)
11. Egenhofer, M.: A mathematical framework for the definition of topological relations. In: International Symposium on Spatial Data Handing, pp. 803–813 (1990)
12. Ferreira, K.R., Camara, G., Monteiro, A.M.V.: An algebra for spatiotemporal data: from observations to events. Trans. GIS **18**(2), 253–269 (2014)
13. Frihida, A., Zheni, D., Claramunt, C., Ghezala, H.B.: Modeling trajectories: a spatio-temporal data type approach. In: DEXA, pp. 447–451 (2009)

14. Grumbach, S., Rigaux, P., Segoufin, L.: The DEDALE system for complex spatial queries. ACM SIGMOD Rec. **27**(2), 213–224 (1998)
15. Guting, R.H., et al.: SECONDO: an extensible DBMS platform for research prototyping and teaching. In: IEEE ICDE, pp. 1115–1116 (2005)
16. Güting, R.H., et al.: A foundation for representing and querying moving objects. ACM TODS **25**(1), 1–42 (2000)
17. Jensen, S.K., Pedersen, T.B., Thomsen, C.: Time series management systems: a survey. IEEE TKDE **29**(11), 2581–2600 (2017)
18. Kucuk, A., Hamdi, S.M., Aydin, B., Schuh, M.A., Angryk, R.A.: PG-trajectory: a PostgreSQL/PostGIS based data model for spatiotemporal trajectories. In: IEEE BDCloud, pp. 81–88 (2016)
19. Li, R., He, H., Wang, R., Ruan, S., Sui, Y., Bao, J., Zheng, Y.: TrajMesa: a distributed NoSQL storage engine for big trajectory data. In: IEEE ICDE, pp. 2002–2005 (2020)
20. Pelekis, N., Frentzos, E., Giatrakos, N., Theodoridis, Y.: HERMES: a trajectory DB engine for mobility-centric applications. IJKBO **5**(2), 19–41 (2015)
21. Pfoser, D., Jensen, C.S.: Capturing the uncertainty of moving-object representations. In: Güting, R.H., Papadias, D., Lochovsky, F. (eds.) SSD 1999. LNCS, vol. 1651, pp. 111–131. Springer, Heidelberg (1999). https://doi.org/10.1007/3-540-48482-5_9
22. Sadri, R., Zaniolo, C., Zarkesh, A., Adibi, J.: Expressing and optimizing sequence queries in database systems. ACM TODS **29**(2), 282–318 (2004)
23. Schek, H.J., Scholl, M.H.: The relational model with relation-valued attributes. Inf. Syst. **11**(2), 137–147 (1986)
24. Trajcevski, G., Wolfson, O., Zhang, F., Chamberlain, S.: The geometry of uncertainty in moving objects databases. In: Jensen, C.S., Šaltenis, S., Jeffery, K.G., Pokorny, J., Bertino, E., Böhn, K., Jarke, M. (eds.) EDBT 2002. LNCS, vol. 2287, pp. 233–250. Springer, Heidelberg (2002). https://doi.org/10.1007/3-540-45876-X_16
25. Waldmann, U., et al.: 3D-MuPPET: 3D multi-pigeon pose estimation and tracking. Int. J. Comput. Vision 1–18 (2024)
26. Wang, S., Bao, Z., Culpepper, J.S., Cong, G.: A survey on trajectory data management, analytics, and learning. ACM Comp. Surv. **54**(2), 1–36 (2021)
27. Wang, S., Bao, Z., Culpepper, J.S., Xie, Z., Liu, Q., Qin, X.: Torch: a search engine for trajectory data. In: ACM SIGIR, pp. 535–544 (2018)
28. Zhan, L., Zhang, Y., Zhang, W., Wang, X., Lin, X.: Range search on uncertain trajectories. In: ACM CIKM, pp. 921–930 (2015)
29. Zheng, K., Trajcevski, G., Zhou, X., Scheuermann, P.: Probabilistic range queries for uncertain trajectories on road networks. In: EDBT, pp. 283–294 (2011)
30. Zheni, D., Frihida, A., Ghezala, H.B., Claramunt, C.: A semantic approach for the modeling of trajectories in space and time. In: Heuser, C.A., Pernul, G. (eds.) ER 2009. LNCS, vol. 5833, pp. 347–356. Springer, Heidelberg (2009). https://doi.org/10.1007/978-3-642-04947-7_41
31. Zimányi, E., Sakr, M., Lesuisse, A.: MobilityDB: a mobility database based on PostgreSQL and PostGIS. ACM TODS **45**(4), 1–42 (2020)
32. Zlatanova, S.: Topological relationships and their use. In: Shekhar, S., Xiong, H., Zhou, X. (eds.) Encyclopedia of GIS, pp. 1–21. Springer, Cham (2016). https://doi.org/10.1007/978-3-319-23519-6_1548-1

Evaluating a Temporal Relational Algebra Supporting Preferences in Temporal Relational Databases

Luca Anselma[1](✉), Antonella Coviello[1], and Paolo Terenziani[2]

[1] Dipartimento di Informatica, Università degli Studi di Torino, Turin, Italy
luca.anselma@unito.it, antonella.coviello@edu.unito.it
[2] DISIT, Università del Piemonte Orientale, Alessandria, Italy
paolo.terenziani@uniupo.it

Abstract. Though a plethora of different approaches have been devoted to the treatment of time within the relational context, temporal preferences have not been faced yet. In several tasks and applications (e.g., planning, scheduling, workflows, guidelines), the exact time of occurrence of facts is not known (i.e., time is *indeterminate*), and preferences can be assigned to the different temporal possibilities. In a recent paper, we have laid the theoretical basis of the treatment of preferential indeterminate time in temporal relational databases, by proposing a relational representation and a temporal relational algebra to manage it, and studying their theoretical properties (e.g., correctness, reducibility). In this paper, we assess such a theoretical framework in terms of performance evaluation along different dimensions, and study the overhead added to cope with preferences with respect to relational approaches (i) without time, (ii) with exact time, and (iii) with indeterminate time (but no preferences).

Keywords: Performance Evaluation · Temporal relational databases · Temporal relational algebra · Temporal indeterminacy · Preferences

1 Introduction

Most phenomena and data have a temporal dimension, so that DataBases (DBs) have to cope with it. Since the 1990s, researchers have pointed out that time has a special nature, so that it has to be managed in through *dedicated* techniques [17]. For instance, in Section 1 of [17] Snodgrass et al. show illuminating examples showing the complexity of managing even apparently trivial queries (e.g., queries involving joins between relations) in case the *valid time* of tuples has to be considered. Hundreds of *dedicated approaches* have been proposed to cope with different temporal phenomena in *temporal relational databases* (TDBs in the following) [21]. TSQL2, a consensus approach of the TDB scientific community, constitutes a landmark in the area [17]. TSQL2 consists of a temporal data model and a query language, grounded on an explicit semantics (BCDM) and on a temporal algebra. Theoretical properties (e.g., reducibility of TSQL2 algebra to Codd's algebra [8] to grant interoperability with non-temporal DB approaches) have been

J. Tekli et al. (Eds.): ADBIS 2024, LNCS 14918, pp. 32–44, 2024.
https://doi.org/10.1007/978-3-031-70626-4_3

also considered in [17]. Most of the major DBMSs nowadays have temporal support, and the research in the TDB area is still quite vibrant (see, e.g., the survey [13] and the recent Encyclopedia of Database Systems [15]).

Despite such a huge amount of work, several problems have not been fully explored yet. For instance, most TDB approaches focus on the time of occurrence of facts (*valid time* [17]) and/or on the time of their insertion/deletion in the DB (*transaction time* [17]), assuming that both times are exactly known. Unfortunately, in many contexts and tasks (e.g., workflows, guidelines, planning, scheduling), the exact valid time is not known in an exact way (in technical terms, *temporal indeterminacy* occurs [12]).

Let us consider Ex.1.

Ex. 1. Mary's physical therapy must start between March 10 and March 30 and last 4–20 days ■

Notably, Ex. 1 is an implicit way of denoting (if we consider the granularity of days) the $21 \cdot 17 = 357$ precise scenarios/instantiations in Ex. 1' below:

Ex. 1'.
Mary's therapy starts on March 10 and lasts 4 days, or

Mary's therapy starts on March 10 and lasts 5 days, or

....

Mary's therapy starts on March 30 and lasts 20 days. ■

Despite the large number of applications in which the valid time of facts is only known in an approximate way, only few TDB approaches have faced temporal indeterminacy. Some of the most relevant approaches in this context include [2, 3, 6, 7, 10, 12]. However, none of such approaches take into account the fact that, in many contexts, it is important to consider that not all the different scenarios/instantiations of a temporally indeterminate fact have the same preference (i.e., in Dubois et al.'s words [11], a *ranking of the instantiations that are acceptable*).

However, research in areas such as Artificial Intelligence shows that managing preferences is very important in connection with temporal indeterminacy: "***temporal preferences*** *are quantitative preferences that pertain to the position and duration of events in time*" [16].

Temporal preferences play a key role for the **personalization** of approaches, and have an important role in many areas, including, e.g., economics [9] and medicine [1, 20]. For instance, in the clinical guideline context, it is not possible/useful to state precisely when actions have to be executed [19], so that temporal constraints in the guidelines can be typically interpreted as general recommendations, to be respected with different levels of medical preferences [1]. Consider, for instance, Ex. 2.

Ex. 2. Mary's physical therapy must start between March 5 and March 13 and last for 6–8 days with high preference, start between March 4 and March 14 and last for 5–13 days with a medium preference, and start between March 1 and March 21 and last 4–20 days with a low preference. ■

In a recent paper, we have laid the theoretical basis of the first framework supporting preferential indeterminate time in TDBs [4]. We have proposed a relational data model coping with preferential indeterminate time, and a temporal relational algebra to query it, and we have studied the theoretical properties (e.g., data semantics, correctness, and reducibility of the algebra). In this paper, we assess such a theoretical framework in terms of performance evaluation. We analyze the performance of our approach along different dimensions and study the added overhead of dealing with preferences.

The paper is organized as follows. To make the paper self-contained, in Sect. 2 we briefly summarize the main results in [4], with specific attention to the aspects related to the performance analysis carried on in this paper. In Sect. 3 we briefly sketch our implementation of the theoretical framework and analyze the performance of our temporal algebraic operators considering sample relations of different dimensions, and different scales of preferences. We also compare the performance of our approach with respect to a reference approach in which the explicit semantics of the data model is managed (such a sematic approach is presented in Sect. 2). Moreover, we analyze the overhead introduced by our treatment of temporal preferences with respect to relational approaches (i) without time, (ii) with exact time, and (iii) with indeterminate time but no preferences. Finally, Sect. 4 presents conclusions and future work.

2 Data Model and Algebra for Temporal Preferences

In [4] we have proposed the first TDB model and relational algebra coping with ***temporal indeterminacy with preferences***. Our approach grounds on the general methodology proposed in [6]:

(1) we provided a compact **representation** (data) **model**,
(2) we provided a **relational algebra** to query it,
(3) we defined the **data semantics** of the model,
(4) we proved the **correctness** of the algebra,
(5) we proved the **reducibility** of our approach to Codd's standard relational algebra.

In the following, to make this paper self-contained, we resume points 1 (in Sect. 2.1) and 2 (in Sect. 2.2), while points 3, 4, and 5 are briefly sketched in Sect. 2.3.

2.1 Data Model

Concerning the temporal domain, as in TSQL2 [17], BCDM [14] and in many approaches reviewed in [17], in our approach time is discrete, linearly ordered, and isomorphic to a subset of the integers. The chronon is the basic time unit. We denote by T^C the chronon domain; in this paper we assume days as chronons.

Concerning preferences, we consider qualitative (i.e., non-numeric) preferences. For generality, we do not commit to a specific scale of preferences (e.g., <high, medium, low> in Ex. 2 above): any scale of qualitative or quantitative preferences is supported as long as a total order can be defined on its elements.

Events, properties, and facts (henceforth we adopt ***facts*** as a cover term for all of them) are associated with the time when they occur (i.e., ***valid time*** [17]), and the relative preferences.

For each preference level (e.g., "high", "medium", "low" in Ex. 2) we specify the range of times having (at least) such a preference, in terms of a pair $< [s_m, s_M), [d_m, d_M)$

> such that $s_m, s_M \in T^C$, $s_m \le s_M$, $d_m, d_M \in N$, $d_m \le d_M$. s_m represents the minimum starting chronon, s_M the maximum starting chronon, d_m the minimum duration, and d_M the maximum duration.

In [4], we focus on a specific class of temporal preference functions, ***layered preference functions*** [5]. Indeed, in many areas and applications, preferences are "centered" on a given set of temporal values and decrease while getting far from this center, forming a sort of "***pyramid***" of *nested* ranges, where the top range has the highest preference and the bottom range the lowest one.

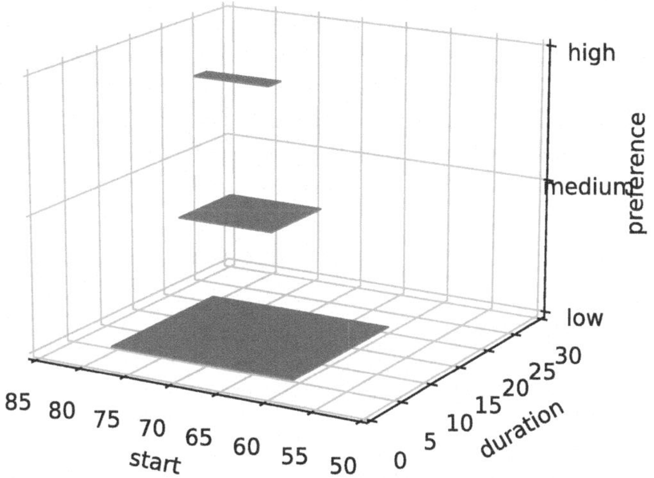

Fig. 1. Pyramid preferences for Ex. 2.

For instance, Fig. 1 shows the tri-dimensional pyramid of preferences corresponding to Ex. 2. In the example, we assume that time starts at the beginning of the year, and we adopt the granularity of days. Thus, March 5 corresponds to day 64, and March 13 to day 72, so that the top rectangle, with coordinates (64, 73, 6, 9), represents a starting time between day 64 and day 73 (*start* axis) and a duration between 6 and 9 days (*duration* axis) when Mary's physical therapy can be performed with high preference (*p* axis). Notably, the rectangle with low preference (coordinates (60, 81, 4, 21)) includes the rectangle with medium preference (coordinates (63, 74, 5, 14)), which includes the rectangle with high preference.

A ***temporal relation with layered preferences*** associates facts with valid times with layered preferences. Given a scale with *r* levels (which is unique for a given DB), the ***schema*** of a ***temporal relation with layered preferences*** $R = (A_1, \ldots ,A_n|(T_{sm}{}^1, T_{sM}{}^1, T_{dm}{}^1, T_{dM}{}^1),\ldots, (T_{sm}{}^r, T_{sM}{}^r, T_{dm}{}^r, T_{dM}{}^r))$ consists of an arbitrary number of non-temporal attributes $A_1, \ldots, A_n$, encoding some fact, and of *r* quadruples $(T_{sm}{}^i, T_{sM}{}^i, T_{dm}{}^i, T_{dM}{}^i)$ of temporal attributes to encode $<[s_m, s_M), [d_m, d_M)>$.

Ex. 3. Ex. 2 can be represented, using the scale <high, medium, low>, by the relation pat_treatm^{TP} with schema $(patient\_t, treatment \mid (T_{sm}{}^1, T_{sM}{}^1, T_{dm}{}^1, T_{dM}{}^1), (T_{sm}{}^2, T_{sM}{}^2, T_{dm}{}^2, T_{dM}{}^2), (T_{sm}{}^3, T_{sM}{}^3, T_{dm}{}^3, T_{dM}{}^3))$ as following:

$$pat\_treatm^{TP} = \{(\text{Mary, phys} \mid (60, 81, 4, 21), (63, 74, 5, 14), (64, 73, 6, 9))\}.$$

2.2 Relational Algebra

A large majority of TDB approaches follows the convention that temporal relational algebra operators behave like standard Codd's operators on the non-temporal attributes, and perform difference (for temporal difference $-^{\mathrm{T}}$), and intersection (for temporal Cartesian product $\times^{\mathrm{T}}$) on the temporal attributes while temporal algebraic operators for selection, projection and union is usually the standard Codd's definition [15, 17, 18]. Our approach follows such a "consensus" background. The formal definition of our temporal algebraic operators is quite complex, especially in the case of temporal difference, and is reported in [4]. Despite this, the idea is quite simple, and has a graphical explanation that clarifies the main concept: temporal Cartesian product has to perform the intersection between two pyramids and temporal difference has to perform difference between pyramids (see Fig. 2 as an example). In the rest of the paper, we use the superscript "TP", temporal preference, for our operators.

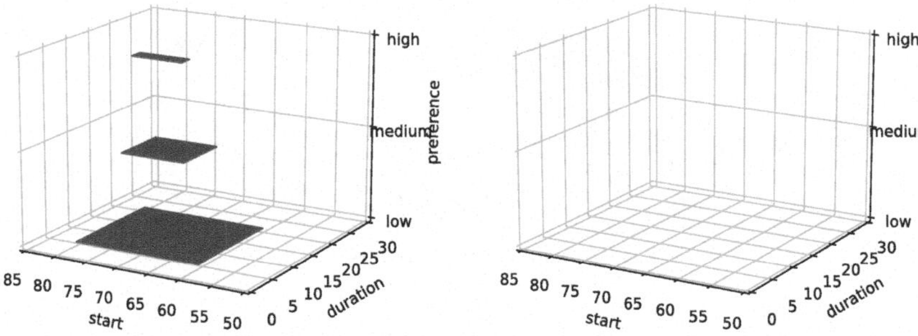

Fig. 2. Temporal Cartesian product where the result is the intersection of the two pyramids, i.e. the orange pyramid (on the left). Temporal difference where the result is the difference between pyramids, i.e. the red pyramids not overlapping with the black one (on the right) (see [4]).

2.3 Data Semantics, Correctness and Reducibility

We propose a compact and implicit representation of the association of valid times and preferences with facts, since four rectangle vertices are used to summarize all the bi-temporal points in each rectangle (see Fig. 1). *Following the methodology* in [6] we have defined the semantics of our data model through the definition of the function *Make-Explicit*, that makes explicit all the pieces of information that we compactly represent in our model.

Specifically, *Make-Explicit*, applied to a temporal relation in our approach, provides as output a new relation in which each fact is paired with the explicit set of all the triples *<s, d, p>* such that *x*, starting at *s* and lasting *d*, has at least preference *p*.

Ex. 4. Make-Explicit(*pat_treatm*) = {(Mary, phys | {<69, 4, L>, ..., <89, 20, L>, <72, 5, M>, ..., <82, 13, M>, <73, 6, H>, ..., <81, 8, H>})}.

(for the sake of brevity, here we have used L, M, and H to denote low, medium, and high respectively). ■

Since our algebraic operators perform a manipulation of an implicit representation, a proof of correctness is required. Informally speaking, for each algebraic operator $\mathrm{Op}^{\mathrm{TP}}$ in our algebra we have proved that, if we first apply $\mathrm{Op}^{\mathrm{TP}}$ to relations in our representation and then we move to an explicit representation through the Make-Explicit function, we get the same result that we would obtain by moving from our representation to an explicit one, and then applying the "explicit" operator $\mathrm{Op}^{\mathrm{Expl}}$ corresponding to $\mathrm{Op}^{\mathrm{TP}}$.

Reducibility is fundamental for all TDB approaches to grant that the new operators, which extend simpler operators to cope with new phenomena, reduce to simpler operators when the new phenomena are disregarded [14, 17]. In [4] we proved such a fundamental property for our algebra.

3 Experimental Evaluation

We have implemented the data model and algebra discussed in Sect. 2 (and the TSQL2 model and algebra, as an example of approach coping with determinate time), using PostgreSQL and PL/pgSQL stored procedures. We have experimentally evaluated the performance of our relational algebra. We performed the experiments on an Intel Core i7-10750H CPU, 2.60 GHZ, 32 GB RAM, OS Ubuntu 24.04, using PostgreSQL 16 DBMS with default settings (effective cache size 500 MB, 8 KB page size and 500 MB shared buffers).

We have generated datasets of different sizes, to test and compare the scalability of our approach. We focused our experiments on the algebraic operators that manipulate the temporal and preference attributes, i.e., Cartesian product and difference.

Both for Cartesian product and for difference, we consider tables with just two non-temporal (text) attributes. Notably, this is a very unfavorable setting for our approach with respect to a situation in which many more non-temporal attributes are considered, since we add to each tuple as many temporal attributes as 4 times the number of preference levels.

For evaluating the *Cartesian product* T1 $\times^{TP}$ T2 we have generated two tables, T1 and T2, of increasing sizes (500, 1000, 2500, 5000 tuples), increasing number of preference levels (3, 5, 10 preference levels) and increasing size of the basis of pyramids (for the sake of simplicity, we assume $s_M - s_m = d_M - d_m$ and we call such a difference *interval size*), such that 10% of the tuples of T1 intersects with their value-equivalent counterpart (i.e., tuples with the same values for the non-temporal attributes) in T2.

For evaluating the *difference* T1 $-^{TP}$ T2, we have generated two tables T1 and T2 of increasing sizes (10000, 25000, 50000, 100000, 250000, 500000 tuples) and increasing number of preference level (3, 5, 10 preference levels), in such a way that 10% of the tuples in T1 have a value-equivalent counterpart in T2. Among them, 20% temporally intersect.

Regarding the measurements of I/O blocks, they have been conducted using the PostgreSQL extension *pg_stat_kcache*. This extension allows for the monitoring and analysis of physical disk access, providing valuable insights into the Input/Output (I/O) operations done by the filesystem layer. Each experiment was run 5 times, and we report average results.

3.1 Experimental Evaluation of the Approach

In this section we experimentally evaluate our approach (indicated by "*Pyramid*" for short) considering different dimensions: first we measure the impact of the number of preference levels, then we compare our approach with an equivalent but explicit approach with different preferences and intervals.

Varying the Number of Preference Levels. We explored the effects of varying the number of preference levels on difference and Cartesian product by utilizing the previously described tables T1 and T2, with *interval size* 10. Figure 3 and 4 showcase a comparative analysis that examines average execution times and I/O operations across varying dataset sizes. These figures specifically highlight how performance metrics are influenced by the application of 3, 5, and 10 preference levels. The data presented aims to elucidate the relationship between preference levels and system efficiency, offering insights into the scalability and resource demands of different preference levels (even if in a realistic setting 5 preference levels might suffice).

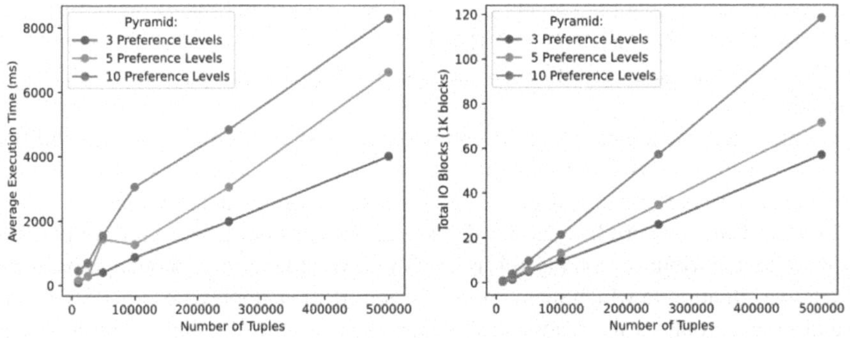

Fig. 3. Difference. Comparisons between varying preference levels as regards execution time (on the left) and I/O (on the right), considering datasets of different sizes.

For both operations, a higher number of preference levels results in increased execution times and I/O blocks. This increase can be attributed to the fact that a higher number of preference levels has a greater number of quadruples and pairs *<start-time, duration>* representing them. For example, managing 10 preference levels could effectively involve processing 10 times the amount of information compared to handling data without any specified preferences. Interestingly, despite the notable increase in complexity with a higher number of preference levels, the rise in execution time and I/O is not directly proportional to the increase in the number of preference levels (e.g., in the case of difference, operating on 10 levels does not lead to 10 times the I/O obtained when operating on one level). This observation suggests that, while the overhead induced by

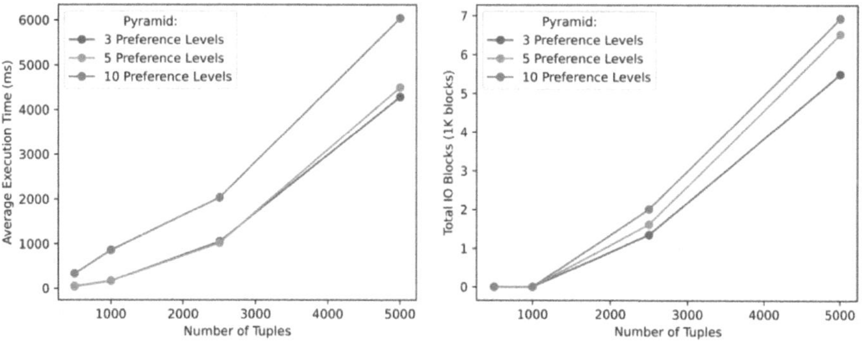

Fig. 4. Cartesian product. Comparisons between varying preference levels as regards execution time (on the left) and I/O (on the right), considering datasets of different sizes.

a higher numbers of preference levels is evident, it is relatively modest considering the steep increase in data complexity.

Comparison with the Explicit Approach. We now compare two approaches to handling temporal indeterminate intervals: our compact representation, referred to as the "implicit approach", and the contrasting "explicit approach". For the explicit approach, we explicitly represented the temporal indeterminate intervals (see the discussion about data semantics and *Make-Explicit* in Sect. 2.3 above).

In the implicit approach, the tables are T1 and T2 as discussed above, with *interval size* 10. In the explicit version, the corresponding tables T1' and T2' contain identical tuples to T1 and T2 with respect to non-temporal attributes and have explicit values for temporal attributes.

This comparison aims to highlight the distinctions in data representation and efficiency between the implicit and explicit approaches, underscoring the potential impact on processing performance and data manageability.

Varying the Number of Preference Levels. Figures 5 and 6 compare the performance of implicit and explicit approaches in handling varying preference levels across difference and Cartesian product focusing on execution time and I/O operations.

The explicit approach consistently exhibits higher execution times and number of I/O blocks compared to the implicit approach across all preference levels. This difference is particularly pronounced in the Cartesian product where the explicit approach's execution time and I/O blocks rapidly rise as the dataset size increases. This suggests that managing preferences explicitly is more computationally and resource-intensive, which may not scale well with larger datasets.

Furthermore, the implicit approach shows a more moderate increase in both execution time and number of I/O blocks, indicating better scalability and efficiency. This is especially relevant in scenarios involving large volumes of data, where the implicit approach could offer substantial performance benefits by handling preferences more efficiently.

Varying Interval Sizes. Figure 7 and 8 show the impact of using different interval sizes in implicit and explicit approaches on both number of I/O blocks and execution

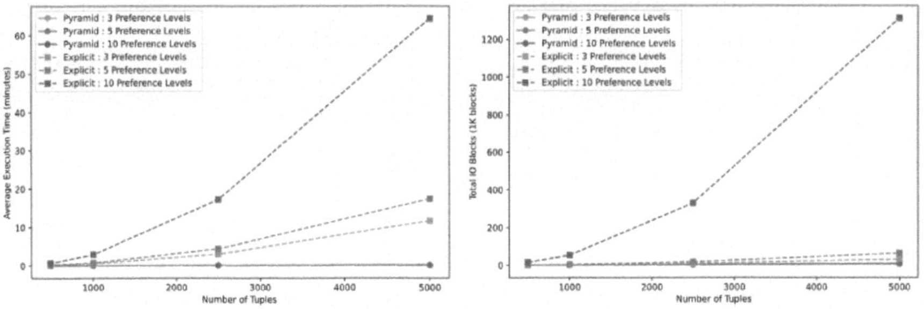

Fig. 5. Cartesian product. Comparisons between implicit and explicit approach with varying number of preference levels as regards execution time (on the left) and I/O (on the right), considering datasets of different sizes.

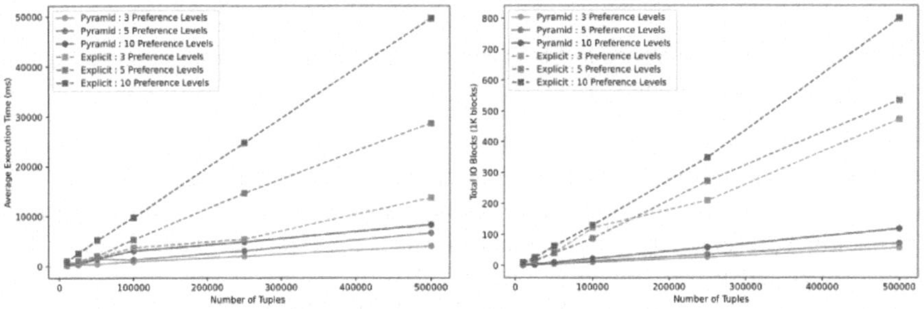

Fig. 6. Difference. Comparisons between implicit and explicit approach with varying number of preference levels as regards execution time (on the left) and I/O (on the right), considering datasets of different sizes.

time when handling data, considering three preference levels. These figures are key to understanding how the choice between implicit and explicit approaches can influence performance across different *interval sizes*, particularly in the context of Cartesian product. The interval sizes tested are 10, 50, and 100. All experiments were capped at a maximum duration of one hour. This time limit has led to the censoring of results for the *interval size* 50 with 5000 tuples, and the *interval size* 100 for datasets larger than 1000 tuples in the explicit approach, as depicted in Fig. 7. The decision to censor these data points in the explicit approach was made to maintain the integrity of the experimental analysis while acknowledging the performance limitations encountered under these specific conditions.

The results indicate that, in the implicit approach, both execution time and I/O operations are not affected by the interval size. Therefore, in Fig. 7 and 8, only the line corresponding to interval size 100 –solid green line– is shown.

In contrast, the explicit approach manifests a considerable increase in both execution time and I/O operations with the increase of interval sizes. This means that the explicit approach incurs a substantial burden on the system when handling larger intervals. This can be explained due to the bigger amount of data that the explicit approach requires to represent triples <*start, duration, preference*>. The pronounced overhead of the explicit

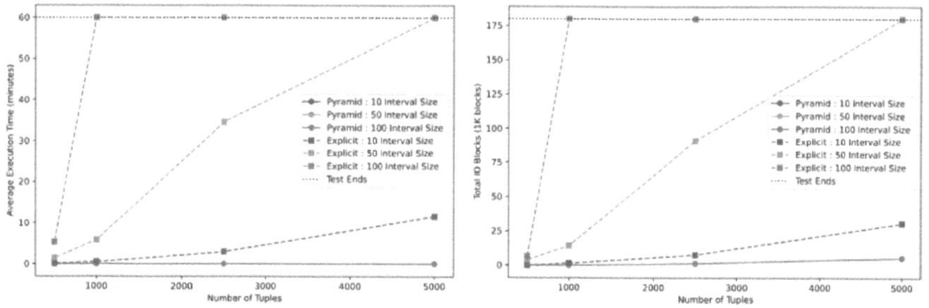

Fig. 7. Cartesian product. Comparisons between implicit and explicit approach with different interval size as regards execution time (on the left) and I/O (on the right), considering datasets of different sizes.

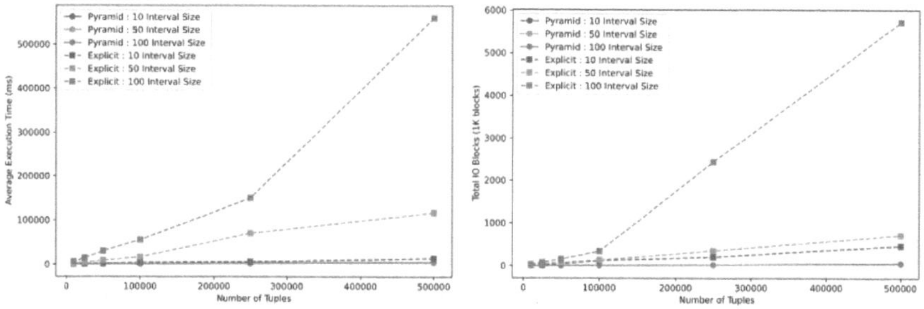

Fig. 8. Difference. Comparisons between implicit and explicit approach with different interval size as regards execution time (on the left) and I/O (on the right), considering datasets of different sizes.

approach to interval size modification suggests potential scalability issues, particularly in contexts involving voluminous datasets or when the data necessitates the management of extensive intervals.

This difference in overhead shows the ability of the implicit method in adapting to larger data intervals while maintaining performance stability, therefore highlighting the implicit approach's better adaptability and minimal impact on system resources even under escalating demands.

3.2 Experimental Evaluation of the Overhead Introduced Compared with Less Expressive Approaches

We now compare our approach with less expressive methods, examining the trade-offs associated with the additional expressiveness provided by our approach. This comparison highlights the overhead incurred when incorporating *indeterminate time* and *preferences*, relative to approaches that lack one or both characteristics.

In the following, we compare four approaches: *Pyramid, Indet, Exact,* and *No-Time.*

Pyramid is our approach with three preference levels.

Indet is an approach with temporal indeterminacy but without preferences; for the sake of simplicity, we implement Indet as the instantiation of our approach using a single preference level.

Exact lacks both indeterminate intervals and preference levels but still incorporates time, i.e., deals with exact time; for this approach, we implemented the algebraic operators of the TSQL2 approach. In our analysis of TSQL2, we implemented the tables T1" and T2" (corresponding to T1 and T2 respectively) such that they have two temporal attributes modelling the start and the end of the valid time and 10% of the tuples of T1" and T2" temporally intersects.

No-Time is meant to explore the absence of both time and preferences, and it corresponds to non-temporal Codd's operators [8]. For this approach we considered a non-temporal variant of the implementation above in which the tables contain the same tuples as T1 and T2 but have only the non-temporal attributes.

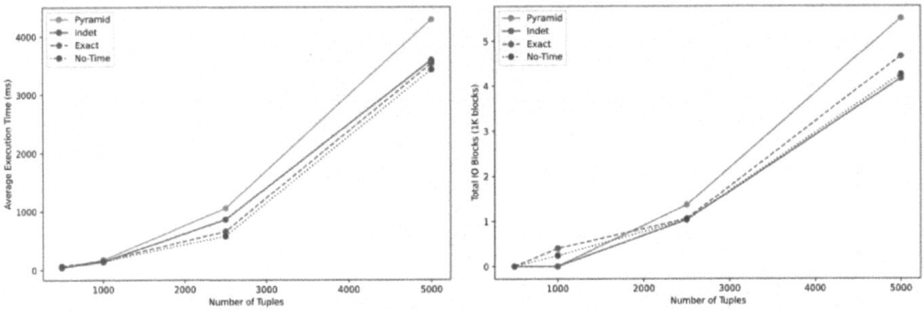

Fig. 9. Cartesian product. Comparisons between implicit approach, TSQL2 and non-temporal with different preference levels as regards execution time (on the left) and I/O (on the right), considering datasets of different sizes.

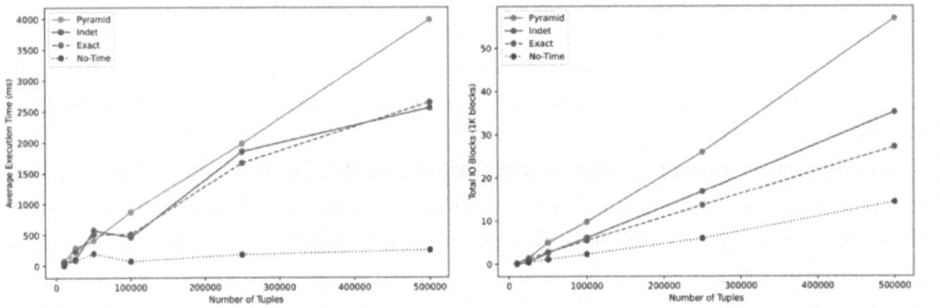

Fig. 10. Difference. Comparisons between implicit approach, TSQL2 and non-temporal with different preference levels as regards execution time (on the left) and I/O (on the right), considering datasets of different sizes.

The experiments reported in Fig. 9 show that the addition of exact time, indeterminate time and of preferences adds a negligible overhead with respect to the non-temporal approach for Cartesian product. Indeed, most of the complexity of Cartesian product both in execution time and I/O depends on the combination of each pair of tuples in the input table. The overhead related to the treatment of preferences is quite limited, and mostly due to the fact that tuples have an increased size (e.g., 12 additional attributes, in the case of three preference levels), and additional interval intersection operations have to be performed.

The experiments concerning difference (shown in Fig. 10) mainly confirm the results we obtained considering Cartesian product. However, in the case of difference, some discrepancies between the performance of the four approaches emerge. In particular, non-temporal difference is significantly more efficient than temporal differences, both as regards execution time and I/O. As regard execution time, no operation on temporal attributes has to be performed. As regards I/O, the size of input tuples is much smaller (no temporal attribute, with respect to the 12 temporal attributes required to deal with three preference levels; notably, in the experiments we consider just two non-temporal attributes), and the output of temporal difference is larger, since also value-equivalent tuples can contribute to it. Moreover, the experiments show that the behavior of the Exact and Indet approaches are very similar, while an additional limited overhead has to be considered concerning our Pyramid approach, due to the increased number of temporal attributes, and, thus, to the additional number of operations between them.

4 Conclusions

In several tasks and applications, the exact time of occurrence of facts is not known (i.e., time is *indeterminate*), and *preferences* can be assigned to the different temporal possibilities. In a recent paper, we have laid the theoretical basis of the first TDB approach supporting preferential indeterminate time. In this paper, we have evaluated the performance of such a framework along different dimensions and studied the overhead to cope with preferences with respect to relational approaches (i) without time, (ii) with exact time, and (iii) with indeterminate time but no preferences. Overall, the experimental evaluations demonstrate that the introduction of multiple preference levels, while adding expressiveness, does not significantly increase overhead, indicating that the addition of indeterminate time management capabilities adds minimal additional cost but significantly enhances the representational capacity of the data.

In our future work we plan to consider all the steps needed to propose a full extension of PostgreSQL to manage temporal preferences exploiting the implementation of temporal algebraic operators used for the performance evaluation discussed in this paper. The initial step involves extending SQL to enable the creation and querying of tables that include temporal preferences. Following the methodology highlighted in Section 24 of the TSQL2 book [17], we will then consider extending the data dictionary, DDL and Query compilers, run-time evaluator, transaction and data manager, possibly exploiting the extensions already provided in https://github.com/scalegenius/pg_bitemporal to manage (bi)temporal tables in PostgreSQL.

Acknowledgments. This study was partially funded by Fondazione Compagnia San Paolo and Fondazione Cassa Depositi e Prestiti (LEAP-AI project grant number D13C23001280007).

Disclosure of Interests. The authors have no competing interests to declare that are relevant to the content of this article.

References

1. Andolina, A., Guazzone, M., Piovesan, L., Terenziani, P.: Temporal reasoning and query answering with preferences and probabilities for medical decision support. Expert Syst. Appl. **195**, 116565 (2022)
2. Anselma, L., Piovesan, L., Sattar, A., Stantic, B., Terenziani, P.: A comprehensive approach to 'now' in temporal relational databases: Semantics and representation. IEEE Trans. Knowl. Data Eng. **28**(10), 2538–2551 (2016)
3. Anselma, L., Piovesan, L., Terenziani, P.: A 1NF temporal relational model and algebra coping with valid-time temporal indeterminacy. J. Intell. Inf. Syst. **47**, 345–374 (2016)
4. Anselma, L., Terenziani, P.: Preferences in temporal relational databases. IEEE Access **12**, 65418–65427 (2024). https://doi.org/10.1109/ACCESS.2024.3397724
5. Anselma, L., Mazzei, A., Piovesan, L., Terenziani, P.: Reasoning and querying bounds on differences with layered preferences. Int. J. Intell. Syst. **36**(5), 1998–2035 (2021)
6. Anselma, L., Piovesan, L., Terenziani, P.: Dealing with temporal indeterminacy in relational databases: an AI methodology. AI Commun. **32**(3), 207–221 (2019)
7. Anselma, L., Terenziani, P., Snodgrass, R.T.: Valid-time indeterminacy in temporal relational databases: semantics and representations. IEEE Trans. Knowl. Data Eng. **25**(12), 2880–2894 (2013)
8. Codd, E.F.: Relational completeness of data base sublanguages. In: Rustin, R. (ed.) Database Systems, pp. 65–98. Prentice Hall and IBM Research Report RJ 987, San Jose (1972)
9. Cohen, J., Ericson, K.M., Laibson, D., White, J.M.: Measuring time preferences. J. Econ. Lit. **58**(2), 299–347 (2020)
10. Dekhtyar, A., Ross, R., Subrahmanian, V.S.: Probabilistic temporal databases, I: algebra. ACM Trans. Database Syst. **26**(1), 41–95 (2001)
11. Dubois, D., Fargier, H., Prade, H.: Possibility theory in constraint satisfaction problems: handling priority, preference and uncertainty. Appl. Intell. **6**(4), 287–309 (1996)
12. Dyreson, C.E., Snodgrass, R.T.: Supporting valid-time indeterminacy. ACM Trans. Database Syst. **23**(1), 1–57 (1998)
13. Gamper, J., Ceccarello, M., Dignös, A.: What's new in temporal databases?. In: Chiusano, S., Cerquitelli, T., Wrembel, R. (eds) ADBIS 2022. LNCS, vol. 13389, pp. 45–58. Springer, Cham (2022). https://doi.org/10.1007/978-3-031-15740-0_5
14. Jensen, C.S., Snodgrass, R.T.: Semantics of time-varying information. Inf. Syst. **21**(4), 311–352 (1996)
15. Liu, L., Tamer Özsu, M.: Encyclopedia of Database Systems, 2nd edn. Springer, Heidelberg (2018). https://doi.org/10.1007/978-1-4614-8265-9. ISBN 978-1-4614-8266-6
16. Rossi, F., Venable, K.B., Walsh, T.: A Short introduction to preferences: between AI and social choice. In: Synthesis Lectures on Artificial Intelligence and Machine Learning, 1st edn. Morgan & Claypool Publishers (2011)
17. Snodgrass, R.T. (ed.): The TSQL2 Temporal Query Language. Kluwer (1995)
18. Tansel, A.U.: Temporal algebras. In: Liu, L., Özsu, M.T. (eds.) Encyclopedia of Database Systems, 2nd edn. Springer, Heidelberg (2018). ISBN 978-1-4614-8266-6
19. Terenziani, P., German, E., Shahar, Y.: The temporal aspects of clinical guidelines. In: Computer-based Medical Guidelines and Protocols, pp. 81–100 (2008)
20. van der Pol, M., Hennessy, D., Manns, B.: The role of time and risk preferences in adherence to physician advice on health behavior change. Eur. J. Health Econ. **18**(3), 373–386 (2017)
21. Wu, Y., Jajodia, S., Sean Wang, X.: Temporal database bibliography update. Temp. Databases Dagstuhl 338–366 (1997)

PG-FD: Mapping Functional Dependencies to the Future Property Graph Schema Standard

Maude Manouvrier(✉) and Khalid Belhajjame

Université Paris-Dauphine, PSL Research University CNRS UMR [7243] LAMSADE, Paris, France
{maude.manouvrier,khalid.belhajjame}@dauphine.fr

Abstract. There has been a notable effort, in the past five years, to develop and promote Graph Query Language, GQL, as a standard for querying graph data, akin to the role SQL plays in relational data querying. Although this goal is still a work in progress, the graph database community has been advancing by not only defining the GQL specification but also introducing additional specifications such as Property Graph Key, PG-Key, and later Property Graph Schema, PG-Schema, for specifying graph schema and dependencies. In this regard, a number of proposals have been made in the literature for expressing Functional Dependencies in graph data. Our first contribution is a survey of such proposals, highlighting important ones and their differences. Our second contribution is a solution for translating dependencies defined within such proposals into dependencies that conform to PG-Schema. Our solution is accompanied by a working prototype for translating graph dependencies into PG-Schema compliant dependencies.

Keywords: Property graphs · Graph dependencies · Unified formalism · Graph schema standard

1 Introduction

Graph databases have gained momentum in recent years and are increasignly used to model data and knowledge, as suggested by the popularity and adoption of graph database systems[1] such as Neo4J, TigerGraph and Amazon Neptune. They are used in many different areas like social networks, recommendation systems, biomedical research, and fraud detection – see in [6,34]. The growing demand for graph database systems has highlighted a significant issue: the presence of heterogeneous data models and query languages. This heterogeneity poses challenges for achieving interoperability among different systems and

[1] https://neo4j.com, https://www.tigergraph.com and https://aws.amazon.com/neptune.

This work received support from the National Research Agency under the France 2030 program, with reference to ANR-22-PESN-0007.

J. Tekli et al. (Eds.): ADBIS 2024, LNCS 14918, pp. 45–59, 2024.
https://doi.org/10.1007/978-3-031-70626-4_4

creates difficulties for users and developers who need to navigate diverse data models and understand the nuances of various query languages: openCypher (Neo4j) [20], GSQL (TigerGraph) [10], Gremlin (Apache TinkerPop) [25] and PGQL (Oracle) [24] for property graphs and, SPARQL[2] for RDF graphs, to cite a few – see in [7] for a review.

The above issues have been recognized by academics and industry alike, and some working groups have started working on standardizing query languages for property graph databases. A notable effort in this respect is GQL (Graph Query Language) [19], where the objective is to achieve a success similar to that of SQL by establishing a new standard for querying graphs. GQL has been officially published as an ISO/IEC standard in April 2024[3].

Within the GQL standardization efforts, several initiatives are currently underway at the time of writing. One of these initiatives is the formalism to define keys for property graphs, called PG-Keys [3], and more recently the specification of language for property graphs schema, PG-Schema [4].

Several approaches are particularly interested in integrity constraints and dependencies for graphs defining, for example, Graph Functional Dependencies (GFD) [18], some of them being temporal [2] or probabilistic [35], Graph Entity Dependencies (GED) [15,16], and recently normal forms for property graphs [29] – see in [7,31] for a review.

With the emergence of PG-Keys [3] and its extension PG-Schema [4], that should lead to future property graph schema standard, we argue that it is essential to understand how existing integrity constraints and dependencies defined for graphs can be expressed using such future standard. This article contributes in the following two ways: (1) it provides a survey of existing data models and types of constraints utilized for graph-shaped data, highlighting their differences and significance and (2) it presents a mapping for translating a notable subset of the identified constraint types into the future property graph schema standard PG-Schema.

The paper is organized as follows. Section 2 contains a brief guide of concepts and notation. Section 3 includes a concise review of related work on Graph Dependencies. Section 4 presents the mapping for translating such dependencies into PG-Schema. Section 5 presents our proof-of-concept prototype. Finally, we conclude and briefly comment on future work in Sect. 6.

2 Preliminaries

Using the same notations and concepts of [7,16,27], let a $\mathcal{O}$ be a set of objects, $\mathcal{L}$ be a finite set of labels, $\mathcal{K}$ be a set of property keys, and $\mathcal{N}$ be a set of values. A *Labeled Property Graph* is a graph $G = (V, E, \eta, \lambda, v)$ where:

- $V \subseteq \mathcal{O}$ is a final set of vertices and $E \subseteq \mathcal{O}$ is a final set of edges;
- $\eta : E \rightarrow V \times V$ is a function assigning an ordered pair of vertices to each edge;

[2] https://www.w3.org/TR/rdf-sparql-query/.

[3] https://www.iso.org/standard/76120.html and https://www.gqlstandards.org/home.

- $\lambda : (V \cup E) \rightarrow \mathcal{P}(\mathcal{L})$ is a function assigning to each object a finite set of labels (i.e., $\mathcal{P}(L)$ denotes the set of finite subsets of set $\mathcal{L}$);
- $v \rightarrow (V \cup E) \times \mathcal{K} \rightarrow \mathcal{N}$ is a partial function assigning values for properties to objects, such that the object sets V and E are disjoint (i.e., $V \cap E = \emptyset$) and the set of domain values, where v is defined, is finite.

The left-hand side of Fig. 1 shows a property graph example representing the current article and its authors, where $V = \{1, 2, 3, 6\}$, $E = \{4, 5, 7\}$, $\lambda(1) = Article$, $\lambda(2) = \lambda(3) = Author$, $\lambda(4) = \lambda(5) = AuthorOf$, $\lambda(6) = Conf$, $\lambda(7) = SubmittedTo$, $\eta(4) = (1, 2)$, $\eta(5) = (1, 3)$, $\eta(7) = (1, 6)$, $v(1, Title) = PG\_FD$, $v(6, Acronym) = ADBIS$, and $v(6, Year) = 2024$, for example.

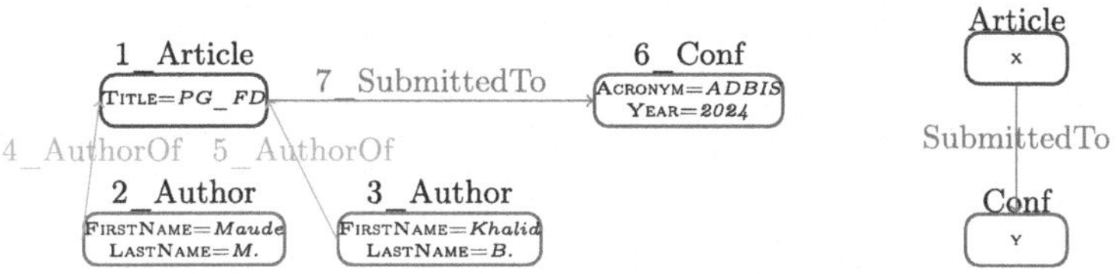

Fig. 1. A Property Graph example (left) and a Graph Pattern example (right)

A *Graph Pattern* is a directed graph $Q[\bar{x}] = (V_Q, E_Q, L_Q)$, where (1) V_Q (E_Q, respectively) is a finite set of pattern nodes (edges, respectively); (2) L_Q is a function that assigns a label to each node $u \in V_Q$ (edge $e \in E_Q$, respectively); and (3) $\bar{x}$ is a list of distinct variables, each denoting a node in V_Q.

An example of a graph pattern is represented in the right-hand side of Fig. 1, where $\bar{x}$ contains two variables: a variable x of label *Article* and a variable y of label $Conf$. Both variables are connected with an edge of label *SubmittedTo*.

A *Functional Dependency* (FD) has been recognized as integrity constraints in databases [8]. In relational databases, a functional dependency $X \rightarrow Y$ is defined on a relational schema R with attribute sets X and Y, where R specifies the "scope" of the FD, meaning that, for each instance r of R and for any tuples t_1 and $t_2 \in r$, if t_1 and t_2 have the same value for X, then t_1 and t_2 must have the same value on Y. A review of functional dependencies in databases can be found in [1,31,32]. In relational databases, the scope of a *relational Functional Dependency* (RFD) spans the entire relation R on which it has been defined, while applied to graph databases, a *Graph FD* (GFD) is defined on each subgraph matching a graph pattern Q [7].

3 Graph Dependencies: Related Work

In this section, we analyze and compare the main types of dependencies proposed in the context of graph data over the last few years, taking into consideration the kind of graph data model the dependency operates on, the method by which the dependency's scope is delineated, and the actual definition of the dependency itself.

Most of the dependencies proposed in the context of graphs are drawn from dependencies proposed in the context of relational databases, in particular functional dependencies. In relational databases, however, the scope of the functional dependency is well delimited intrinsically by the relation to which the attributes participating in the FD belong. This is not the case with graph data. As such, in the context of graph databases, in addition to the dependency itself, information about the scope of the dependency delimiting the sub-graphs in which the dependency is valid, is necessary.

Table 1. Acronyms and references of related work

Acronym	Definition	References
gFD	Graph-tailored functional dependency	[29]
GKs	Graph Keys	[12]
CFDs	Conditional Functional Dependencies	[13]
GPAR	Graph-Pattern Association Rule	[17]
GFD	Graph Functional Dependency	[13]
TGFD	Temporal Graph Functional Dependency	[2]
GED	Graph Entity Dependency	[15,16]
GDC	Graph Denial Constraint	[16]
GAR	Graph Association Rules	[14]
GD	Graph Dependency	[36]
GDD	Graph Differential Dependency	[21]
GPD	Graph Probabilistic Dependency	[35]

Table 1 presents the references of graph dependencies approaches, and Fig. 2 recalls the relationships between all theses approaches. A part of the related work define graph dependencies based on graph pattern (See definition in Sect. 2). These approaches are presented in Sect. 3.1. We particularly focus on the ones of W. Fan's research Team [11] who has defined the GED approach [16] that has been widely extended (see red node in Fig. 2). In Sect. 3.2, we present Graph-tailored Functional Dependencies (gFD) of [29] that targets normalization of graph data. Section 3.3 finally presents how constraints are defined in PG-Schema [4], the unifying language for property graph, which serves as a recommendation for GQL standard.

3.1 Pattern-Based Graph Dependencies

Keys are a special case of FDs. Keys for graphs (GKs), identifying entities in RDF graph, have been defined in [12]. In this work, keys are interpreted by means of graph pattern matching via sub-graph isomorphism. Analogous to GKs, *Graph-Pattern Association Rules* GPARs [17] also capture graph patterns to define constraints on graph data, by studying association rules between entities [31].

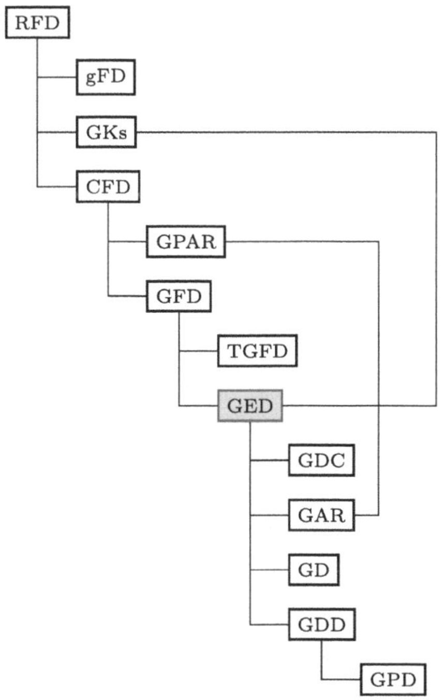

Fig. 2. The relationships among graph dependencies related work

Subsuming Functional Dependencies (FDs) and Conditional[4] Functional Dependencies (CFDs) of [13], *Graph Functional Dependencies* (GFD) have been defined in [18]. A GFD consists of a graph pattern, identifying entities on which the dependency is defined, and an extension of CFDs specifying the dependencies of the attribute values of the entities. *Temporal Graph Functional Dependencies* (TGFDs) of [2] extend GFD to temporal graph (i.e. sequence of graph snapshots).

Later, *Graph Entity Dependencies* (GEDs) [15,16] have been defined to extend GFDs by supporting literal equality of entity id over two nodes in the graph pattern [2]. GED subsumes GFD of [18] and GKs of [12] – see in [11] for a complete comparison of GKs, GFD and GED.

As defined in [7,31], a GED φ is a pair $Q[\bar{x}](X \rightarrow Y)$ where: $Q[\bar{x}]$ is a graph pattern (See definition in Sect. 2) and, $X \rightarrow Y$ is a FD to be applied to entities identified by Q, such that X and Y are two (possibly empty) sets of literals of $\bar{x}$. A literal of $\bar{x}$ can be:

- a constant literal $x.A = c$ where $A \in \mathcal{K}$ is an attribute of $x \in \bar{x}$ and $c \in \mathcal{N}$ is a constant, meaning $v(x, A) = c$;
- a boolean constant $False$, meaning that pattern $Q[\bar{x}]$ is "illegal";
- a variable literal $x.A = y.B$ where $A, B \in \mathcal{K}$ are attributes of the respective entities $x, y \in \bar{x}$, meaning $v(x, A) = v(y, B)$;

[4] i.e. FDs that conditionally hold in a part of the relation [31].

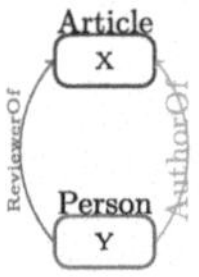

$\varphi_1 = \mathcal{Q}_1[x, y](\emptyset \rightarrow False)$

$\mathcal{Q}_1[x, y]$

```
FOR (x:Article)
 COUNT 0 OF (y:Person)-[:AuthorOf]->(x)<-[:ReviewerOf]-(y:Person)
```

Fig. 3. An "illegal" pattern GED and its translation into PG-Schema

- an id literal $id(x) = id(y)$, $x, y \in \bar{x}$ and $id()$ denoting the vertex or edge identities. In this case, pattern $Q[\bar{x}]$ is composed of two similar sub-patterns.

Y can be interpreted as the disjunction of its literals, in an extension of GED called GED$^{\vee}$s – see in [16] for more details.

The top of Fig. 3 represents an example of a GED, where X is empty and $Y = False$, meaning that the pattern $Q[x, y]$ is "illegal". Indeed, an author of an article cannot be a reviewer of one of its articles. The top of Fig. 4 represents an other GED example, used to repair data. It means that if two *Article* entities x and x' have the same title and are published in the same conference then, both articles are the same. An article is therefore identified by its title and the identifier of the conference in which it was published.

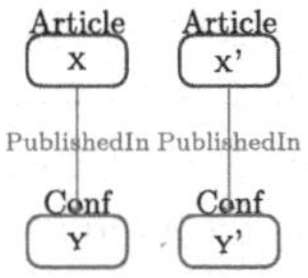

$\varphi_2 = \mathcal{Q}_2[x, x', y, y'](X_2 \rightarrow Y_2)$
with $X_2 = \{x.title = x'.title,\ y.id = y'.id\}$
and $Y_2 = \{x.id = x'.id\}$

$\mathcal{Q}_2[x, x', y, y']$

```
FOR (x:Article) IDENTIFIER x.title, y.id WITHIN (x)-[:PublishedIn]->(y:Conf)
```

Fig. 4. A vertex identity GED and its translation into PG-Schema

GEDs have been extended to *Graph Differential Dependencies* (GDDs) in [21], by incorporating distance and matching function. *Graph Probabilistic Dependencies* (GPDs) [35] extend GDD with probabilistic and statistical approaches in order to relax the dependency constraints. *Graph Association Rules* (GARs) extend GFDs and GEDs with the existential semantics for attributes and edges, and by allowing ML (machine learning) classifiers as predicates [14]. *Graph Denial Constraints* (GDCs) [16] extend GEDs, replacing the equality relationship in the definition of GEDs with built-in predicates $=$, $\neq$, $<$, $>$, $\leq$ and $\geq$ [31]. Recent approaches have been proposed in [22,37] to discover GEDs in property graphs.

As explained in [29], the aforementioned approaches define graph dependencies that compare values of properties or constants for all pairs of entities identified by a graph pattern. The next section presents other graph dependencies having a different expressiveness requiring the existence of some graph objects, i.e. nodes, edges or properties.

3.2 Existence-Based Functional Dependency

Several approaches limit the graph objects on which the graph dependencies hold by existence conditions.

Recently, the *graph-tailored functional dependency* (gFD), defined in [29], restricts the set of vertices on which the graph dependency holds to the nodes for which all properties in a property set P exist, P containing at least all the properties that occur in the dependency. More formally, a gFD is defined as an expression $L : P : X \to Y$ where $L \subseteq \mathcal{L}$ and $X, Y \subseteq P \subseteq \mathcal{K}$. $L : P : X \to Y$ is satisfied iff there are no vertices $v_1, v_2 \in V$ such that $v_1 \neq v_2$, for all $A \in P$, $v(v_1, A)$ and $v(v_2, A)$ are defined, for all $A \in X$, $v(v_1, A) = v(v_2, A)$ and for some $A \in Y$, $v(v_1, A) \neq v(v_2, A)$. Basically, a gFD indicates that the combination of values on some node properties uniquely determine the values on some other node properties.

For example, in the property graph of Fig. 1, we can define the gFD stipulating that $Conf$ nodes, with properties *Acronym* and $ConfName$, have matching values on $ConfName$ whenever they have matching values on *Acronym*:

$$Conf : \{Acronym, ConfName\} : ConfName \to Acronym$$

The same authors also defined, in [28,30], *graph-tailored uniqueness constraint* (gUC), which is an expression $L : P : X$, which is satisfied iff there are no vertices $v_1, v_2 \in V$ such that $v_1 \neq v_2$, for all $A \in P$, $v(v_1, A)$ and $v(v_2, A)$ are defined, for all $A \in X$, $v(v_1, A) = v(v_2, A)$. For example, we have:

$$Conf : \{Acronym, ConfName, Year\} : ConfName, Year$$

As explained in [29], every gUC $L : P : X$ implies the gFD $L : P : X \to P$. According to the authors, *Neo4j* UCs of [23] are gUCs [29], which are a sub-class of PG-Keys [3], that we explain in the next section.

The Graph Dependency (GD) of [36] extends GED[5] of [15,16] with existence constraint. A GD φ is a pair $Q[\bar{x}](X \to Y)$ where $Q[\bar{x}]$ is a topological pattern in the knowledge graph, and X and Y are extended from the GED definition (see Sect. 3.1):

- X is extended to the existence of a graph object (node or edge) in the knowledge graph with certain properties value or labels which can be constant or

[5] Authors of [36] said they adapt GFD of [18]. However, they used graph database and a syntax inspired from *Cypher* language [20] and used node id. Therefore, they rather extend GED than GFD.

related to $\bar{x}$. More formally, X is expressed as an existing condition, such as $\exists o \in V$ or $\exists o \in E$, associated with predicates such as $o.label = \ell$ with $\ell \in \mathcal{L}$, or $o.A = c$ with c a constant and $A \in \mathcal{K}$, or $o.A = x.A$ with $x \in \bar{x}$.
- Y is extended to support connection between nodes defined in X and nodes in $\bar{x}$. Y is expressed in ASCII art notation of Cypher [20], also adopted by GQL [19]: `(x)->(y)` means that there exists an edge between nodes x and y and `(x)-[e]->(y)` states that edge between nodes x and y is e.

For example, using the Graph Pattern $Q[x, y]$ of the right-hand side of Fig. 1, we can define a constraint indicating that all *Article* nodes should be connected to a *Conf* one with the following GD: $Q[x, y]$, $X \rightarrow Y$, with X is $\exists$ edge $e \in E$, $\lambda(e) = SubmittedTo$ and Y is `(x)-[e]->(y)`.

3.3 PG-Schema

PG-Schema [4] specifies a schema language for property graphs that should lead to future schema standards [19]. A constraint in PG-Schema, that extends the work on keys for property graphs [3], called PG-Keys, is of the form:

$$\texttt{FOR}\, p(x) <qualifier> q(x, \bar{y})$$

where both $p(x)$ and $q(x, \bar{y})$ are queries, respectively called the scope and the descriptor. $<qualifier>$ consists of combinations of `EXCLUSIVE`, `MANDATORY`, `SINGLETON`, `IDENTIFIER` and `COUNT LB..UB OF ...`, where LB an integer representing a lower bound and UB is an (optional) integer representing an upper bound. Statements in PG-Schema constraint are of the form:

- `FOR` $p(x)$ `MANDATORY` $q(x, \bar{y})$ means that for every output x of $p(x)$ there should be at least one tuple $\bar{y} = (y_1, y_2, .., y_n)$ that satisfies $q(x, \bar{y})$. As explained in [28], `MANDATORY` combines existence and uniqueness constraints.
- `FOR` $p(x)$ `SINGLETON` $q(x, \bar{y})$ means that for every output x of $p(x)$ there should be at most one tuple $\bar{y} = (y_1, y_2, .., y_n)$ that satisfies $q(x, \bar{y})$.
- `FOR` $p(x)$ `EXCLUSIVE` $q(x, \bar{y})$ means that no $\bar{y}$ should be shared by two different values of x. As explained in [28], `EXCLUSIVE` separates existence and uniqueness constraints.
- `FOR` $p(x)$ `IDENTIFIER` $q(x, \bar{y})$ is equivalent to: `FOR` $p(x)$ `EXCLUSIVE MANDATORY SINGLETON` $q(x, \bar{y})$.
- `FOR` $p(x)$ `COUNT LB..UB OF` $q(x, \bar{y})$ expresses that the number of distinct results returned by $q(x, \bar{y})$ must be between the lower bound and the optional upper bound, whereas $p(x)$ `COUNT 0 OF` $q(x, \bar{y})$ means that $q(x, \bar{y})$ returns an empty set.

For example, to express the constraint indicating that "an article should at least have one author" using PG-Schema, we can write:

```
FOR x:Article MANDATORY e,y WITHIN (x)-[e:AuthorOf]->(y:Author)
```

Keyword `WITHIN` in PG-Schema specifies the pattern to be matched in the property graph.

As explained in [4], PG-Schema allows to handle key and foreign key constraints defined in property graph and can also express express SQL-style CHECK constraints and SQL-style `CHECK` constraints or denial constraints.

4 Expressing Graph Dependency Through PG-Schema

We show, in this section, how we translate graph dependencies into property graph constraints using the PG-Schema language defined by [4]. In Subsect. 4.1, we focus on *Graph Entity Dependencies* (GEDs) of [16], which is described in Sect. 3.1. Subsection 4.2 is dedicated to gFD and gUC of [29], described in Sect. 3.2. Subsection 4.3 refers to GD [36], presented in Sect. 3.2. Finally, Subsect. 4.4 presents how we translate relational FD using PG-Schema, as relational databases can be converted into graphs (see in [9] for example). A proof of the soundness of our proposed translation is presented in Subsect. 4.5.

4.1 From GED to PG-Schema

In GED [16], $\mathcal{Q}[\bar{x}]$ represents a Graph Pattern where $\bar{x}$ is a list of distinct variables, each one denoting a vertex. Let x be the first variable of $\bar{x}$ and L its label ($\lambda(x) = L$). Variable x corresponds to the node on which the dependency is defined.

The following rules (numbered from 1 to 4) represent the translation into PG-Schema of GEDs according to the different forms of FD $X \rightarrow Y$. `Q`, in PG-Schema constraint, represents the pattern $\mathcal{Q}[\bar{x}]$ using the ASCII art: `(var1:NodeLabel)-[var2:EdgeLabel]->(var3:NodeLabel)` with $var1$, $var2$ and $var3$ variables that can be optional.

1. When X and Y only consist of constant literals:

```
FOR (x:L) WHERE X MANDATORY Y WITHIN Q
```

2. When X is $\emptyset$ and Y consists of variable literals:

```
FOR (x:L) MANDATORY Y WITHIN Q
```

3. When X consists of variable literals and Y consists of id literals:

```
FOR (x:L) IDENTIFIER left side of variable literals X WITHIN Q'
```

with `Q'` *a sub-pattern of* `Q` *– An example of this form of FD is given in Fig.* 4, *where pattern* `Q` *consists of two article vertices linked to two conference vertices.* `Q'` *represents one article vertex and its related conference vertex.*

4. When X is $\emptyset$ and Y is $False$:

```
FOR (x:L) COUNT 0 OF Q
```

An example of this form of FD is given in Fig. 3 *detailed in Sect.* 4.1.

For example, the bottom of Fig. 3 represents the translation of GED $\mathcal{Q}_1$ (above in the figure) into PG-Schema. It means that, for a node x of label *Article*, there does not exists any node y of label *Person* participating in both *AuthorOf* and *ReviewerOf* relations. The bottom of Fig. 4 represents the translation of GED $\mathcal{Q}_2$, meaning that an article is identified by its title and the id of the conference where it has been submitted.

4.2 From gFDs to PG-Schema

As explained in [28,29], a gUC $\{L_1, ..., L_m\} : \{P_1, ..., P_n\} : \{U_1, ..., U_k\}$, with $\forall i$, $L_i \in \mathcal{L}$, $P_i \in \mathcal{K}$ and $\{U_1, ..., U_k\} \subset \{P_1, ..., P_n\}$ can be specified using PG-Schema by the following constraint:

```
FOR x:L1 ... :Lm WHERE x.P1 IS NOT NULL ... AND x.Pn IS NOT NULL
 EXCLUSIVE x.U1, ..., x.Uk
```

For example, the gUC, $Conf$: $\{Acronym, ConfName, Year\}$: $ConfName, Year$ can be translated to:

```
FOR x:Conf WHERE x.Acronym IS NOT NULL AND x.ConfName IS NOT NULL
                                       AND x.YEAR IS NOT NULL
 EXCLUSIVE x.ConfName, x.Year
```

A gFD [29], defined by $\{L_1, ..., L_m\} : \{P_1, ..., P_n\} : \{X_1, ..., X_k\} \rightarrow \{Y_1, ..., Y_j\}$ where $\{L_1, ..., L_m\} \subseteq \mathcal{L}$ and $\{X_1, ..., X_k\}, \{Y_1, ..., Y_j\} \subseteq \{P_1, ..., P_n\} \subseteq \mathcal{K}$, can therefore be translated using the `EXCLUSIVE MANDATORY` of PG-Schema:

```
FOR x.X1, ..., x.Xk WITHIN x:L1 ... :Lm
  WHERE x.X1 IS NOT NULL AND ... AND x.Xk IS NOT NULL
   EXCLUSIVE MANDATORY x.Y1, ..., x.Yj
```

For example, the gFD $Conf$: $\{Acronym, ConfName\}$: $ConfName \rightarrow Acronym$ can be expressed into PG-Schema by:

```
FOR x.ConfName WITHIN x:Conf
  WHERE x.ConfName IS NOT NULL AND x.Acronym IS NOT NULL
    EXCLUSIVE MANDATORY x.Acronym
```

4.3 From GD to PG-Schema

The following two rules represent the translation into PG-Schema of GD, proposed in [36] and presented in Sect. 3.2.

1. When X is $\exists e \in E$, $\lambda(e) = L$ and Y is `(x)-[e]->(y)`:

```
FOR (x:x.label) MANDATORY e,y WITHIN (x)-[e:L]->(y:y.label)
```

2. When X is $\exists x \in V$, $\lambda(x) = L$ and Y is `(x)->(y)`:

```
FOR (x:L) MANDATORY e,y WITHIN (x)-[e]->(y:y.label)
```

For instance, the example of GD presented in Sect. 3.2, using the Graph Pattern $Q[x, y]$ of Fig. 1 (right-hand side), and $X \rightarrow Y$, with X defined by $\exists$ edge $e \in E$, $\lambda(e) = SubmittedTo$ and Y is defined by `(x)-[e]->(y)`, can be translated using PG-Schema by:

```
FOR (x:Article) MANDATORY e,y WITHIN (x)-[e:SubmittedTo]->(y:Conf)
```

4.4 From Relational Databases to PG-Schema

Relational databases can be converted to property graph [9] or into RDF store [26] that can be mapped into property graph [5].

Therefore, a functional dependency (FD) $X \rightarrow Y$ is defined on a relation schema $R(\mathcal{K}_R)$, with $\mathcal{K}_R$ the attribute set containing sets $X = \{X_1, X_2, .., X_n\}$ and $Y = \{Y_1, Y_2, .., Y_m\}$, can also be translated into PG-Schema by:

```
FOR x.Y1, ... Y.Ym WITHIN (x:R) EXCLUSIVE MANDATORY x.X1, ..., X.Xn
```

meaning that the same value of X cannot be shared by two values of Y.

For example, the FD $address \rightarrow region$, with $address$ and $region$ being two attributes of a schema R, means that each address corresponds to an explicit region. When relation R is mapped to a vertex R, the aforementioned FD is translated to:

```
FOR x.region WITHIN (x:R) EXCLUSIVE MANDATORY x.address
```

4.5 Proof of the Soundness of the Translation

In this section, we outline the sketch of the proof that can be used to demonstrate the soundness of the translation operations presented thus far. Let $\mathcal{M}$ be a graph data model, which could encompass various existing proposals ranging from relational data models to triple-based graph models (RDF) and property graph models. Consider a data dependency *dep* expressed within the data model $\mathcal{M}$. Note that PG-Schema considers graph data expressed using property graphs, which is an expressive data model that encompasses all the data models used by existing proposals. Consequently, an instance $\mathcal{I}$ of a data model $\mathcal{M}$ can be translated to a property-graph instance $\mathcal{I}^{\mathcal{PG}}$ without any loss of information.

As defined in [5], the properties of a mapping are: computability, semantics preservation and information preservation. Our translation is computable. Indeed, it is easy to see that the mapping rules we define can be implemented as an algorithm. An implementation is presented in the next section. Our translation is also information preserving because, due to our mapping rules, it does not lose any information about the graph dependency being translated. Indeed, we can define inverse rules to recover the original dependency from the PG-Schema constraint resulting from the translation process.

To establish the soundness of our proposed translation for transforming a dependency *dep* expressed in a model $\mathcal{M}$ into a dependency dep^{PS} compliant with PG-Schema, and then prove its semantics preservation, it suffices to prove that:

$$\forall \mathcal{I} \in instances(\mathcal{M})\ I \models dep \implies I^{PG} \models dep^{PS} \quad (1)$$
$$I \not\models dep \implies I^{PG} \not\models dep^{PS} \quad (2)$$

In what follows, we show the case where the source model $\mathcal{M}$ is the relational model. Other cases can be proven in a similar manner, e.g. [15] showing that relational FDs are special cases of GEDs, when tuples in a relation are represented as nodes in a graph.

Consider the following FD: $X \rightarrow Y$ with $X = \{X_1, X_2, .., X_n\}$ and $Y = \{Y_1, Y_2, .., Y_m\}$. According to the solution presented in the previous section, the corresponding PG-Schema dependency is expressed as follows:
`FOR x.Y1, ... Y.Ym WITHIN (x:R) EXCLUSIVE MANDATORY x.X1, ..., X.Xn.`

Consider an instance $\mathcal{I}$ of the relational model where FD holds, signifying that any two tuples t_1 and t_2 sharing identical attribute values in X also share identical values in Y. By contradiction, we can demonstrate that the PG-Schema dependency mentioned earlier also applies to the instance $\mathcal{I}^{\mathcal{PG}}$, thereby showing (1).

Now, suppose FD does not hold in a given relational model $\mathcal{I}$, indicating the existence of two tuples t_1 and t_2 sharing the same values for attributes in X but differing in at least one attribute Y_i in Y. Upon translation of $\mathcal{I}$ into $\mathcal{I}^{\mathcal{PG}}$, two vertices representing t_1 and t_2 emerge, each associated with attributes or properties (edges) specifying values for attributes in X and Y. In this scenario, the PG-Schema dependency defined previously does not hold, as the two vertices in question are linked to the same values (or vertices) in X but not to the same attribute (or vertex) in Y. Hence, this shows (2).

5 Proof-of-Concept Prototype Implementation

To prove that the PG-schema mapping we propose in this article is computable, we have developed a Python prototype called *PG-FD*. In order to manage graph pattern, for GED of [16] and GD of [36], our prototype uses the `network`[6] Python module, which is a powerful library for manipulating property graphs in Python. To manage functional dependencies, for gFDs of [29] or relational schema for example, *PG-FD* uses the Python module functional_dependencies[7], that defines the classe *FD* to represent a functional dependency.

To translate GED (resp. GD) into PG-Schema, user should call function `GED2PGS(GP,X,Y)` (resp. `GD2PGS(GP,X,Y)`), with `GP` the graph pattern, and `X` and `Y` the right and left part of the graph dependency (coded by list of strings). To define the graph pattern, user can either manually define it using the `MultiDiGraph`[8] class of the `network` module or using *Cypher*/*GQL* ASCII art

[6] https://networkx.org/.
[7] https://oer.gitlab.io/cs/functional-dependencies/.
[8] https://networkx.org/documentation/stable/reference/classes/multidigraph.html.

style, which will be translated into `MultiDiGraph` by our prototype. Our prototype could be extended to allow the user to create a graph in a graph database (e.g. *Neo4j*) or using *PG-Types* of PG-Schema, that describes the shape of data and the types of graph components such as nodes and edges. To translate a gFD of [29] into PG-Schema, user should call function `gFD2PGS(L,P,fd)` with `L` a set of node labels, `P` a set of properties and `fd` a functional dependency defined using the `functional_dependencies` module. This module is also used to translate a relational FD, using function `Rel2PGS(fd,R)`, with `R` the relation name.

The source code of our prototype can be downloaded from GitHub[9]. The prototype can be used to replay the examples of this article, some GED examples defined in [15], as well as to define new graph dependencies to translate.

6 Conclusions

In this article, we have surveyed the different solutions for defining graph functional dependencies in the literature. We have also defined mapping rules allowing to translate graph dependencies, focusing on prominent ones, into the PG-Schema [4], which serves as a recommendation or upcoming versions of GQL for specifying property graph schema. A proof of the soundless of the proposed translation and a proof-of-concept implementation prototype are presented in this article.

In our ongoing research, we are extending our solution to cater for other forms of graph-based dependencies. For example, [33] presents a representation similar to functional dependencies and is used to convert SQL constraints into SHACL (Shapes Constraint Language). We also aim to conduct comprehensive experiments on graph dependencies discovered in real-world graphs.

Acknowledgment. We would like to thank Ibtissem Khedim, for facilitating the initiation of this project during her master's research internship, and the referees of the article for their insightful feedback.

References

1. Abedjan, Z., Golab, L., Naumann, F., et al.: Data Profiling. Springer, Cham (2022). https://doi.org/10.1007/978-3-031-01865-7
2. Alipourlangouri, M.: Temporal dependencies for graphs. In: ACM SIGMOD Conference (2021). https://doi.org/10.1145/3448016.3450586
3. Angles, R., Bonifati, A., Dumbrava, S., et al.: PG-keys: keys for property graphs. In: ACM SIGMOD Conference (2021). https://doi.org/10.1145/3448016.3457561
4. Angles, R., Bonifati, A., Dumbrava, S., et al.: PG-schema: schemas for property graphs. PACMMOD **1**(2), 1–25 (2023). https://doi.org/10.1145/3589778
5. Angles, R., Thakkar, H., Tomaszuk, D.: Mapping RDF databases to property graph databases. IEEE Access **8**, 86091–86110 (2020). https://doi.org/10.1109/ACCESS.2020.2993117

[9] https://github.com/MaudeManouvrier/PG-FD.

6. Boncz, P.: The (sorry) state of graph database systems (2022). Keynote in EDBT
7. Bonifati, A., Fletcher, G., Voigt, H., et al.: Querying graphs, vol. 10, no. 3, pp. 1–184 (2018). https://doi.org/10.1007/978-3-031-01864-0
8. Codd, E.F.: Further normalization of the data base relational model. Data Base Syst. **6**, 33–64 (1972)
9. De Virgilio, R., Maccioni, A., Torlone, R.: Converting relational to graph databases. In: GRADES (2013). https://doi.org/10.1145/2484425.2484426
10. Deutsch, A., Xu, Y., Wu, M., et al.: Aggregation support for modern graph analytics in TigerGraph. In: ACM SIGMOD Conference (2020). https://doi.org/10.1145/3318464.3386144
11. Fan, W.: Dependencies for graphs: challenges and opportunities. ACM J. Data Inf. Qual. **11**(2), 1–12 (2019). https://doi.org/10.1145/3310230
12. Fan, W., Fan, Z., Tian, C., et al.: Keys for graphs. VLDB Endow. **8**(12), 1590–1601 (2015). https://doi.org/10.14778/2824032.2824056
13. Fan, W., Geerts, F., Jia, X., et al.: Conditional functional dependencies for capturing data inconsistencies. ACM TODS **33**(2), 1–48 (2008). https://doi.org/10.1145/1366102.1366103
14. Fan, W., Jin, R., Liu, M., et al.: Capturing associations in graphs. VLDB Endow. **13**(12) (2020). https://doi.org/10.14778/3407790.3407795
15. Fan, W., Lu, P.: Dependencies for graphs. In: ACM SIGMOD-SIGACT-SIGAI Symposium (2017). https://doi.org/10.1145/3034786.3056114
16. Fan, W., Lu, P.: Dependencies for graphs. ACM TODS **44**(2), 1–40 (2019). https://doi.org/10.1145/3287285
17. Fan, W., Wang, X., Wu, Y., et al.: Association rules with graph patterns. VLDB Endow. **8**(12), 1502–1513 (2015). https://doi.org/10.14778/2824032.2824048
18. Fan, W., Wu, Y., Xu, J.: Functional dependencies for graphs. In: ACM SIGMOD Conference (2016). https://doi.org/10.1145/2882903.2915232
19. Francis, N., Gheerbrant, A., Guagliardo, P., et al.: A researcher's digest of GQL. In: ICDT Conference, vol. 255 (2023). https://doi.org/10.4230/LIPIcs.ICDT.2023.1
20. Francis, N., Green, A., Guagliardo, P., et al.: Cypher: an evolving query language for property graphs. In: ACM SIGMOD Conference (2018). https://doi.org/10.1145/3183713.3190657
21. Kwashie, S., Liu, L., Liu, J., et al.: Certus: an effective entity resolution approach with graph differential dependencies (GDDs). VLDB Endow. **12**(6), 653–666 (2019). https://doi.org/10.14778/3311880.3311883
22. Liu, D., Kwashie, S., Zhang, Y., et al.: An efficient approach for discovering graph entity dependencies (GEDs). Inf. Syst. **125** (2024). https://doi.org/10.1016/j.is.2024.102421
23. Pokorný, J., Valenta, M., Kovačič, J.: Integrity constraints in graph databases. Procedia Comput. Sci. **109**, 975–981 (2017). https://doi.org/10.1016/j.procs.2017.05.456
24. van Rest, O., Hong, S., Kim, J., et al.: PGQL: a property graph query language. In: GRADES (2016). https://doi.org/10.1145/2960414.2960421
25. Rodriguez, M.A.: The gremlin graph traversal machine and language (invited talk). In: DBPL (2015). https://doi.org/10.1145/2815072.2815073
26. Sequeda, J.F., Arenas, M., Miranker, D.P.: On directly mapping relational databases to RDF and OWL. In: WWW Conference (2012). https://doi.org/10.1145/2187836.2187924
27. Shimomura, L.C., Fletcher, G., Yakovets, N.: GGDs: graph generating dependencies. In: ACM CIKM (2020). https://doi.org/10.1145/3340531.3412149

28. Skavantzos, P., Leck, U., Zhao, K., et al.: Uniqueness constraints for object stores. ACM J. Data Inf. Qual. (2023). https://doi.org/10.1145/3581758
29. Skavantzos, P., Link, S.: Normalizing property graphs. VLDB Endow. **16**(11), 3031–3043 (2023). https://doi.org/10.14778/3611479.3611506
30. Skavantzos, P., Zhao, K., Link, S.: Uniqueness constraints on property graphs. In: La Rosa, M., Sadiq, S., Teniente, E. (eds.) CAiSE 2021. LNCS, vol. 12751, pp. 280–295. Springer, Cham (2021). https://doi.org/10.1007/978-3-030-79382-1_17
31. Song, S., Chen, L.: Integrity Constraints on Rich Data Types. Springer, Cham (2023). https://doi.org/10.1007/978-3-031-27177-9
32. Song, S., Gao, F., Huang, R., et al.: Data dependencies extended for variety and veracity: a family tree. IEEE TKDE **34**(10), 4717–4736 (2020). https://doi.org/10.1109/TKDE.2020.3046443
33. Thapa, R.B., Giese, M.: Mapping relational database constraints to SHACL. In: Sattler, U., et al. (eds.) ISWC 2022. LNCS, vol. 13489, pp. 214–230. Springer, Cham (2022). https://doi.org/10.1007/978-3-031-19433-7_13
34. Tian, Y.: The world of graph databases from an industry perspective. ACM SIGMOD Rec. **51**(4), 60–67 (2023). https://doi.org/10.1145/3582302.3582320
35. Zada, M.S.H., Yuan, B., Anjum, A., et al.: Large-scale data integration using graph probabilistic dependencies (GPDs). In: IEEE/ACM BDCAT Conference (2020). https://doi.org/10.1109/BDCAT50828.2020.00028
36. Zheng, X., Dasgupta, S., Gupta, A.: P2KG: declarative construction and quality evaluation of knowledge graph from polystores. In: Abelló, A., et al. (eds.) ADBIS 2023. CCIS, vol. 1850, pp. 427–439. Springer, Cham (2023). https://doi.org/10.1007/978-3-031-42941-5_37
37. Zhou, G., Kwashie, S., Zhang, Y., et al.: FASTAGEDS: fast approximate graph entity dependency discovery. In: Zhang, F., Wang, H., Barhamgi, M., Chen, L., Zhou, R. (eds.) WISE 2023. LNCS, vol. 14306, pp. 451–465. Springer, Singapore (2023). https://doi.org/10.1007/978-981-99-7254-8_35

Discovery and Data Analysis

Generating SPARQL Queries for Data Discovery

Zoé Chevallier[1,2(✉)], Zoubida Kedad[1], Béatrice Finance[1], and Frédéric Chaillan[2]

[1] David Laboratory, University of Versailles Paris-Saclay, Versailles, France
{zoe.chevallier,zoubida.kedad,beatrice.finance}@uvsq.fr
[2] Grand Paris Sud, Evry-Courcouronnes, France
{z.chevallier,f.chaillan}@grandparissud.fr

Abstract. The number of datasets available on the Web represents a huge space of meaningful information. When done manually, identifying the relevant portion of the data for a specific need is time-consuming. This paper presents an approach that identifies and extracts the relevant portion of RDF data sources for some application requirements described by a target schema. The identification process is based on a semi-supervised learning algorithm that iteratively computes for each class of the target schema the similarity between source entities and both the class description and the candidate instance patterns already identified for this class. The extraction process generates SPARQL queries for the identified candidate instance patterns. We present some evaluations showing the effectiveness of our approach.

Keywords: Data discovery · Web Data Extraction · Semi-Supervised Learning · SPARQL Query Generation

1 Introduction

The web of data provides access to a huge set of interconnected datasets, expressed in RDF[1], the standard language proposed by the W3C for describing resources on the web. Many applications, such as visualization, simulation, or decision-making tools, can exploit these data sources. However, exploring the available sources to identify the relevant subset for a specific need is time-consuming. If we assume that the requirements of a given application are described by a conceptual schema, referred to as the target schema, the task of finding relevant information in an RDF data source can be stated as finding candidate instances for the target schema in the data source. Instantiating the target schema is not straightforward, as the alignment between the source and the target schema [3] might not be available. In this paper, we presents an approach that determines whether a source contains data which is relevant for a given target schema, by extracting candidate instance patterns. Each pattern is a property set describing candidate instances in the data source. Our approach relies on

[1] https://www.w3.org/RDF/.

J. Tekli et al. (Eds.): ADBIS 2024, LNCS 14918, pp. 63–76, 2024.
https://doi.org/10.1007/978-3-031-70626-4_5

a semi-supervised learning algorithm that iteratively compares the source entities to both the target classes and the candidate instances already identified for these classes. Additionally, this paper proposes an approach to generate a set of SPARQL[2] queries that extract the candidate instances corresponding to the identified patterns.

Our contribution is twofold: first, an approach for identifying candidate instance patterns for a given target schema in an RDF dataset, even when the alignment between the source and the target schemas is not provided; and second, a SPARQL query generation algorithm which returns the extraction queries enabling to populate the target schema from the source considering the identified instance patterns.

The paper is organized as follows. Section 2 introduces the problem statement. Section 3 presents our approach for the identification of the relevant part of a data source. Section 4 describes the generation of the SPARQL queries. Section 5 presents some experiments, Sect. 6 is devoted to the related works, and finally, Sect. 7 concludes the paper.

2 Problem Statement

Let us consider a target schema T composed of a set of classes. Our problem consists, given a class C defined in T and an RDF data source S, (i) to determine if S is contributive to C, i.e., to check if S contains candidate instances for C, and (ii) if S is a contributive source, to generate the SPARQL queries that will extract the identified candidate instances.

An RDF data source is a set of triples $D \subseteq (R \cup B) \times P \times (R \cup B \cup L)$, where R, B, P, and L represent respectively the resources set, the blank-node set, the property set and the literal set. In a triple <s, p, o>, s is a resource called the subject of the triple, p is a property describing the resource s, and o is the object of the triple, representing the value of the property p for the resource s. The object is either a resource or a literal. An RDF data source contains the data and the schema describing these data, e.g., the definition of classes and properties. In our work, we define an *entity* in an RDF data source as a resource that is not a class, a property, a literal, or a blank node. An entity e is characterized by a set of properties, denoted $Prop(e)$ such that: $Prop(e) = \{p| < e, p, * > \in \mathrm{S}\}$. A target schema T is defined using the RDFS[3] and OWL[4] languages. We assume that a set of target classes C_T is defined in T, such that: $C_T = \{C \mid$ <C, rdf:type, rdfs:Class> $\in T$ or <C, rdf:type, owl:Class> $\in T\}$. Let C be a class defined in the target schema T. A class C is characterized by a set of properties defined in T, denoted $Prop(C)$, such that $Prop(C) = \{$p | <C, p, *> $\in$ T$\}$.

When the alignment between the target schema T and a source S is available, it is straightforward to identify the relevant portion of S for T: if a target class C_t and a source class C_s are equivalent, then all the instances of C_s could

[2] https://www.w3.org/TR/rdf-sparql-query/.
[3] https://www.w3.org/TR/rdf-schema/.
[4] https://www.w3.org/TR/owl-features/.

be instances of C_t. However, these correspondences are not always provided. Besides, the schema in RDF sources is only descriptive, and an instance can be characterized by a property set that is different from the one defined in the schema for its class.

Given a target class C and a source S, our problem is to characterize the relevant portion of S to populate C, and to automatically generate the corresponding extraction SPARQL queries. This relevant portion is a set of entities where each entity e is similar to C with respect to a similarity function f. We present our approach in the following sections.

3 Identifying Candidate Instance Descriptions

Given an RDF source S and a target class C, our goal is to determine if S is contributive to the class C, i.e., if this source contains some candidate instances for C. To this end, we search in the data source for the patterns describing candidate instances for each target class C.

When a source class is equivalent to the target class C, then all its instances can be considered as candidate instances for C. Besides, we consider that an entity is likely to be a candidate instance for a class if it shares a significant number of properties with this class, i.e., if $Prop(e)$ and $Prop(C)$ are similar. However the property set characterizing an instance of C can be different from the one defined in the schema for C. Therefore, we also consider that an entity e in S is likely to be an instance of C if its property set is similar to the one of an existing candidate instance of C.

For each class in the target schema, we propose to extract all the patterns, i.e., the distinct property sets describing candidate instances, by computing the similarity between each entity e and both the class description and the identified candidate instance patterns. When a new candidate pattern is discovered, the set used to compute the similarity evolves, which changes the similarity value. This makes the identification of candidate instance patterns an iterative process.

We introduce the notion of candidate instance description to represent the extracted patterns. It is defined as follows.

Definition 1. ***Candidate Instance Descriptions***
Let C be a target class, and S an RDF data source. Let I_C be a set of candidate instances for C. A candidate instance description $\mathcal{D}_i$ is a set of properties such that there is a candidate instance e in I_C for which $\mathcal{D}_i = Prop(e)$.
The set of all the candidate instance descriptions for a class C is denoted $\mathcal{D}(C)$.

To determine if an entity e is a candidate instance for a target class C, we consider two types of similarities: schema-based similarity which compares the property set of C as defined in the schema to the property set of e, and instance-based similarity which compares the property set of e to the candidate instance descriptions already identified for C.

Definition 2. ***Schema-based Similarity***
Schema-based similarity between an entity e and a class C is based on the Jaccard index [7]. The similarity is denoted $Sim_S(e, C)$, and is such that:

$$Sim_S(e, C) = \frac{|Prop(e) \cap Prop(C)|}{|Prop(e) \cup Prop(C)|} \tag{1}$$

Definition 3. ***Instance-based Similarity***
Instance-based similarity between an entity e and a class C associated to a set of candidate descriptions $\mathcal{D}(C)$ is defined as the maximum value of the similarity between e and each candidate description $\mathcal{D}_i$ in $\mathcal{D}(C)$. It is computed based on the Jaccard index [7]. Let $\mathcal{D}(C) = \{\mathcal{D}_i \mid 1 \leq i \leq n\}$ be a set of candidate instance descriptions. Instance-based similarity is denoted $Sim_I(e, \mathcal{D}(C))$, and is such that:

$$Sim_I(e, \mathcal{D}(C)) = MAX_{1 \leq i \leq n}\left(\frac{|Prop(e) \cap \mathcal{D}_i|}{|Prop(e) \cup \mathcal{D}_i|}\right) \tag{2}$$

If the similarity between an entity e and a class C is higher than a similarity threshold denoted τ, i.e., if $Sim(e, C) \geq \tau$, then $Prop(e)$ is a candidate instance description of C. In order to identify candidate instance descriptions for a class C in a source S we propose to adapt an iterative semi-supervised learning algorithm, the Self-Training algorithm [13]. Algorithm 1 describes the identification of candidate descriptions from a data source for a target class C. It uses the similarity measures introduced above.

If some correspondences between the source and the target schema are provided, we compute the candidate instance descriptions based on the schema alignment (lines 2–4). Then, the entities are compared to the property set of C, $Prop(C)$. If the schema-based similarity between e and C is higher than a threshold τ, their property sets are considered as candidate descriptions (lines 5–7). For each entity e, if the similarity between e and the target class C is higher than a threshold τ, i.e., $Sim_S(e, C) > \tau$ then $Prop(e)$ is added to the set of candidate instance descriptions of the current iteration, $\Delta\mathcal{D}$ (lines 11–13).

The set of candidate instance descriptions identified during the iteration is then added to the set of candidate descriptions of C, $\mathcal{D}(C)$ (line 14).

The algorithm iterates as long as new candidate instance descriptions are identified. The process terminates when no new candidate descriptions are discovered during an iteration, as expressed by the condition in lines 15–16.

Using our algorithm, the same set of properties can be a candidate instance description for several classes. Indeed, assume that a candidate instance description $\mathcal{D}_k$ is extracted for a class C_i; if $\mathcal{D}_k$ is compared to another class C_j and has either a schema-based or an instance-based similarity score above the similarity threshold, then $\mathcal{D}_k$ is also a candidate instance description for C_j.

4 Generating SPARQL Queries

In this section, we describe the process of generating the queries allowing to extract the entities which correspond to the candidate instance descriptions identified in a data source. The generated queries are expressed using the SPARQL query language. A *CONSTRUCT* query could be generated for each identified candidate instance description.

Algorithm 1 Identifying Candidate Instance Descriptions for a Target Class C

Input: RDF data source (S), target schema (T), target class (C), similarity threshold (τ)
Output: $\mathcal{D}(C)$ set of candidate descriptions for C

$\mathcal{D}(C) \leftarrow \emptyset$ ▷ *initialization of the candidate instance descriptions (classifier)*
for all C_S equivalent to C **do** ▷ *schema-alignment based*
    for all e instance of C_S **do**
        $\mathcal{D}(C) \leftarrow \mathcal{D}(C) \cup Prop(e)$
for all e entity in S **do** ▷ *schema-based classification*
    if $Sim_S(e, C) \geq \tau$ **then**
        $\mathcal{D}(C) \leftarrow \mathcal{D}(C) \cup Prop(e)$
condition ← True ▷ *iterations until no new candidate descriptions identified*
while condition **do**
    $\Delta\mathcal{D} \leftarrow \emptyset$
    for all e entity of $S \wedge Prop(e) \notin \mathcal{D}(C)$ **do** ▷ *prediction of entity type*
        if $Sim(e, C) \geq \tau$ **then**
            $\Delta\mathcal{D} \leftarrow \Delta\mathcal{D} \cup \{Prop(e)\}$
    $\mathcal{D}(C) \leftarrow \mathcal{D}(C) \cup \Delta\mathcal{D}$ ▷ *update of the classification model*
    if $\Delta\mathcal{D}$ is empty **then** ▷ *checking of the iteration condition*
        condition ← False
Return $\mathcal{D}(C)$

If we consider a candidate instance description $\mathcal{D}_i = \{p_1, ..., p_n\}$, the generated SPARQL query will extract all the entities e such that $Prop(e) = \mathcal{D}_i$, and their associated triples. It is also possible to generate a single query for several candidate descriptions. For example, if $\mathcal{D}_1$ is composed of the properties $\{p_a, p_b, p_c\}$, and another description $\mathcal{D}_2$ is composed of the properties $\{p_a, p_b\}$, we could execute a single query to extract entities with properties p_a and p_b, and optionally p_c.

In the sequel, we describe our query generation process. Section 4.1 details query generation for a single candidate description, Sect. 4.2 presents an approach enabling to group the candidate descriptions for which a single query can be generated, and Sect. 4.3 presents query generation from a set of descriptions.

4.1 Generating a SPARQL Query for a Single Description

Consider a target class C, a data source S and a candidate instance description $\mathcal{D}_i$ of C such that $\mathcal{D}_i = \{p_1, ..., p_n\}$. Our goal is to generate a SPARQL query for $\mathcal{D}_i$ in order to extract all the entities characterized by a property set identical to $\mathcal{D}_i$, i.e., the set in which each entity e is such that $Prop(e) = \mathcal{D}_i$.

The query is composed of two clauses: the clause *CONSTRUCT* contains a triple pattern for each property in the description, and a type declaration for this entity, which is the target class being instantiated. The *WHERE* clause contains a triple pattern for each property, and a filter to extract only the entities having a property set identical to the considered description.

The *CONSTRUCT* clause generates the instances of a target class C and the associated triples. It contains the triple pattern <?e, rdf:type, C>. For each property p in $\mathcal{D}_i$, the triple pattern <?e, p, $?o_p$> is added.

The *WHERE* clause includes a triple pattern <?e, p, $?o_p$> for each property p in $\mathcal{D}_i$. We implement an exclusion mechanism to filter out the entities which are characterised by a superset of $\mathcal{D}_i$. We use a *MINUS* operator in the query in order to exclude any entity which is described by a property that is not contained in the description, i.e., an entity e having the associated triple <?e, ?prop, ?o> such that $?prop \notin \mathcal{D}_i$. Figure 1 depicts the SPARQL query generated for the description: $\mathcal{D}(t{:}Person) = \{\text{firstname, lastname}\}$.

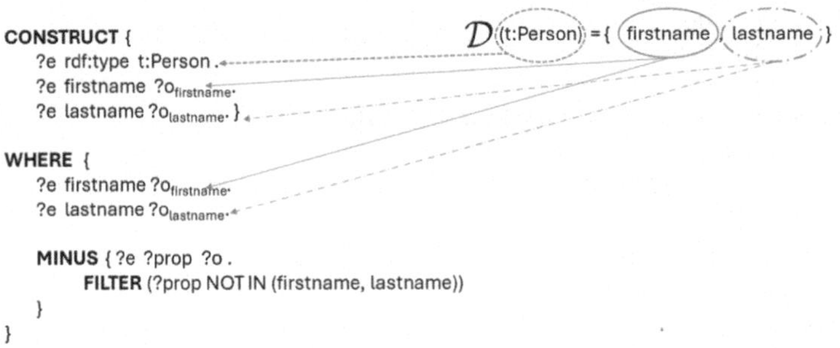

Fig. 1. A Candidate Instance Description and the Corresponding SPARQL Query

4.2 Partitioning the Set of Candidate Instance Descriptions

In order to identify the candidate descriptions that can be processed by a single query, we introduce in this section a partitioning principle.

As an example, two descriptions $\mathcal{D}_1 = \{p_1, p_2\}$, and $\mathcal{D}_2 = \{p_1, p_2, p_3\}$ can be grouped to generate a single query that extracts entities with the properties p_1 and p_2, and optionally p_3. However, consider two descriptions $\mathcal{D}_1 = \{p_1, p_2\}$ and $\mathcal{D}_3 = \{p_1, p_2, p_3, p_4\}$. If we generate a single query that extract entities with the properties p_1, p_2 and optionally p_3, p_4, then an entity characterized by the properties $\{p_1, p_2, p_3\}$ will also be returned by this query, even though $\{p_1, p_2, p_3\}$ is not a candidate description. The goal of this section is to define a partition of the candidate descriptions which groups together the ones that can be processed by the same query. It is defined as follows.

Definition 4. ***Partition of candidate instance descriptions***
A partition $\mathcal{P}(C) = \{\mathcal{P}_1, ..., \mathcal{P}_m\}$ *of the set of candidate instance descriptions* $\mathcal{D}(C)$ *is such that* $\mathcal{D}(C) = \cup_{1 \le i \le m}(\mathcal{P}_i)$, *and for each* $\mathcal{D}_j \in \mathcal{D}(C)$, *if* $\mathcal{D}_j \in \mathcal{P}_i$ *then* $\mathcal{D}_j \notin \mathcal{P}_{k, \forall k \neq i}$.

Considering that $\mathcal{D}_{min}$ and $\mathcal{D}_{max}$ represent the candidate descriptions having respectively the lowest and the highest number of properties in a subset $\mathcal{P}_i$ of $\mathcal{P}(C)$, then every subset of $\mathcal{D}_{max}$ that is also a superset of $\mathcal{D}_{min}$ is in $\mathcal{P}_i$.

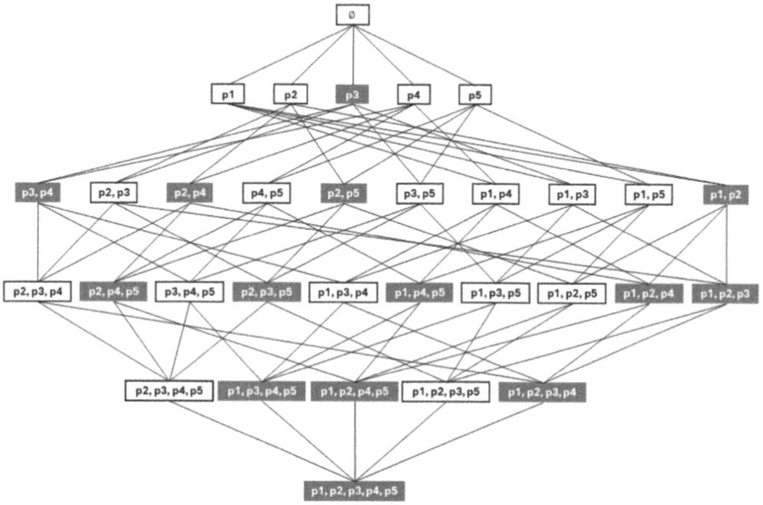

Fig. 2. A Candidate Instance Description Lattice

In order to generate such partition, we represent all the possible combinations of properties as a lattice. Some of the vertices of this lattice represent candidate descriptions in $\mathcal{D}(C)$.

Example. In Fig. 2 the lattice of the possible property combinations for $\mathcal{D}(C)$ is depicted. The candidate instance descriptions are represented as grey rectangles. A possible partition of the candidate instance descriptions is $\mathcal{P}(C) = \{\mathcal{P}_1, \mathcal{P}_2, \mathcal{P}_3, \mathcal{P}_4, \mathcal{P}_5\}$, where $\mathcal{P}_1 = \{\{\text{p3}\}, \{\text{p3, p4}\}\}$, $\mathcal{P}_2 = \{\{\text{p2,p4}\}, \{\text{p2,p4,p5}\}\}$, $\mathcal{P}_3 = \{\{\text{p2,p5}\}, \{\text{p2,p3,p5}\}\}$, $\mathcal{P}_4 = \{\{\text{p1,p2}\},\{\text{p1,p2,p4}\}, \{\text{p1,p2,p3}\}, \{\text{p1,p2, p3, p4}\}\}$, $\mathcal{P}_5 = \{\{\text{p1, p4, p5}\}, \{\text{p1, p2, p4, p5}\}, \{\text{p1,p3, p4, p5}\}, \{\text{p1, p2, p3, p4, p5}\}\}$.

Algorithm 2 details the partitioning process. The vertices are explored in a breadth-first manner from the top of the lattice. For each unmarked vertex V_i corresponding to a candidate description, a subset $\mathcal{P}_k$ is initiated. We then search for the candidate description V_j, a descendent of V_i having the highest number of properties and for which all the vertices in any path from V_i to V_j are unmarked candidate descriptions. Algorithm 3 describes this search. V_j and all the retrieved candidate descriptions are marked and added to $\mathcal{P}_k$.

Algorithm 2 Partitioning the Candidate Instance Descriptions

Input: Λ a complete lattice from properties in $\mathcal{D}(C)$
Output: $\mathcal{P}(C)$ a partition of $\mathcal{D}(C)$

$\mathcal{P}(C) \leftarrow \emptyset$
Q.enqueue(root(Λ))
label root(Λ) as explored
while Q is not empty **do**
  $v :=$ Q.dequeue(
  if CD(v) **then**
    $\mathcal{D}_{min} := v$
    $\mathcal{P}_i := \mathcal{D}_{min}$
    $\mathcal{D}_{max} :=$ maximumCardinalityPartitionSubset($\mathcal{D}_{min}$, labeledVertices)
    for all Vertices b from $\mathcal{D}_{min}$ to $\mathcal{D}_{max}$ **do**
      $\mathcal{P}_i := \mathcal{P}_i \cup \{b\}$
      label b as explored
    $\mathcal{P}(C) := \mathcal{P}(C) \cup \mathcal{P}_i$
  for all Neighbor b of v in Λ **do**
    if b unlabeled as explored **then** Q.enqueue(b)
return $\mathcal{P}(C)$

4.3 Generating a SPARQL Query from a Partition's Subset of Candidate Descriptions

Consider $\mathcal{P}_i$ a subset of the partition $\mathcal{P}(C)$, and consider that $\mathcal{D}_{min}$ and $\mathcal{D}_{max}$ represent the candidate descriptions having respectively the lowest and the highest number of properties in $\mathcal{P}_i$.

The *CONSTRUCT* clause contains a type triple pattern <?e, rdf:type, C> and also the triple patterns $<?e, p, ?o_p>$ for each p property in $\mathcal{D}_{max}$. The *WHERE* clause contains a triple pattern $<?e, p, ?o_p>$ for each property p in $\mathcal{D}_{min}$, and a clause *OPTIONAL* $\{<?e, p, ?o_p>\}$ for each p in $\mathcal{D}_{max} \setminus \mathcal{D}_{min}$. A *MINUS* operator is used to filter out the entities characterized by other properties than the ones in $\mathcal{D}_{max}$.

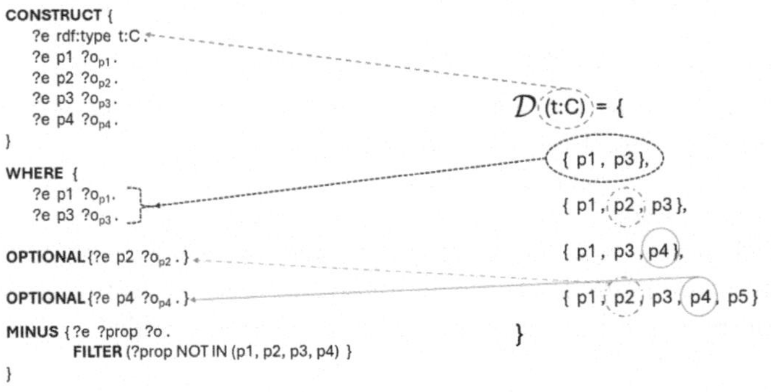

Fig. 3. A Partition Subset and the Corresponding SPARQL Query

Algorithm 3 maximumCardinalityPartitionSubset($\mathcal{D}_i$,labeledNodes)

Input: $\mathcal{D}_{min}$, labeledNodes
Output: $\mathcal{D}_{max}$

Q := leaf(Λ)
label leaf(Λ) as explored
while Q is not empty **do**
    $\mathcal{D}_{max}$:= Q.dequeue(); boolean := True
    for all Vertex V | $pred(\mathcal{D}_{max}) = V$ &$\mathcal{D}_{min} \subset V$ **do**
        if $V \notin \mathcal{D}(C)$ & $V \notin \mathcal{P}(C)$ **then**
            boolean := False
    if boolean **then**
        Return $\mathcal{D}_{max}$
    for all Neighbor b of $\mathcal{D}_{max}$ in Λ **do**
        if {b unlabeled as explored |$\exists$ path from b to $\mathcal{D}_{min}$} **then** Q.enqueue(b)
Return $\mathcal{D}_{min}$

Example. The query for the subset $\mathcal{P}_i$ ={{p1, p3}, {p1, p2, p3}, {p1, p3, p4}, {p1, p2, p3, p4}} is given in Fig. 3.

5 Evaluations

In this section, we will present our experiments. We first introduce the datasets and the methodology, then we present some evaluation results to show the effectiveness of our approach.

5.1 Scenarios and Datasets

Scenario 1. DBpedia and ChatGPT Generated Target Schema. The source instances are extracted from DBpedia[5]. Each dataset corresponds to one class, and we have randomly collected instances of this class with all their associated triples. We have extracted three datasets and presented their characteristics in Table 1. The extracted instances are typed, and the type declarations are considered as the ground truth. We have removed 80% of the type declarations, and 20% of the instances of each class remained typed. This is close to what can be observed on real data sources; for example, according to Fernandez et al. [5], DBpedia contains 16% of typed entities. We have excluded the properties that are part of the RDF, RDFS, or OWL vocabularies as they can be used to describe any class and could introduce a distortion in the result. The target schema has been generated using a large language model, ChatGPT[6]. We have issued a query for a target schema composed of Books, Writers, Publishers

[5] https://www.dbpedia.org/.
[6] https://chat.openai.com/.

and their characteristics. We have assumed the following correspondences: (i) the target class **Book** is equivalent to the source class **Book**, (ii) the target class **Publisher** is equivalent to the source class **Publisher**, (iii) there is no correspondence involving the target class **Writer**.

Table 1. Description of the Datasets

	Number of entities	Number of distinct properties	Average number of properties per entity	Number of distinct patterns
dbo:Book	7572	534	15	5648
dbo:Publisher	2712	534	9	2042
dbo:Writer	9385	737	12	7271
lodib:source1	152905	15	3	14
lodib:source2	155724	17	3	15
lodib:source3	127916	16	4	8

Scenario 2. Linked Open Data Integration Benchmark [11]**.** We have also tested our approach on an existing benchmark. We have considered a target schema and three sources generated by the LODIB tool [11]. The characteristics of the datasets for both scenarios are presented in Table 1. In this table, a high number of distinct patterns indicates a high level of heterogeneity.

5.2 Methodology

We have evaluated our approach by comparing the resulting candidate instance descriptions obtained by both our approach and a baseline approach which relies on schema-based similarity and exploits schema alignment. We have used the recall, the precision, and the f1-score as evaluation metrics.

5.3 Candidate Instance Descriptions Discovered by Our Approach

Figure 4 shows the source classes of the candidate instance descriptions obtained for the target class $tgt{:}Person$ in scenario 2. Figure 4a shows the source types of the candidate instance descriptions, and Fig. 4b shows the evolution of recall, precision, and f1-score.

We can see in Fig. 4a that the baseline identifies candidate instances exclusively from the class Person in source 1 ($src1{:}Person$), while our approach identifies candidate instances from the classes Person in sources 1, 2 and 3 ($src1{:}Person$, $src2{:}Person$, and $src3{:}Person$). Figure 4b shows that iterations 2 and 3 improve the recall.

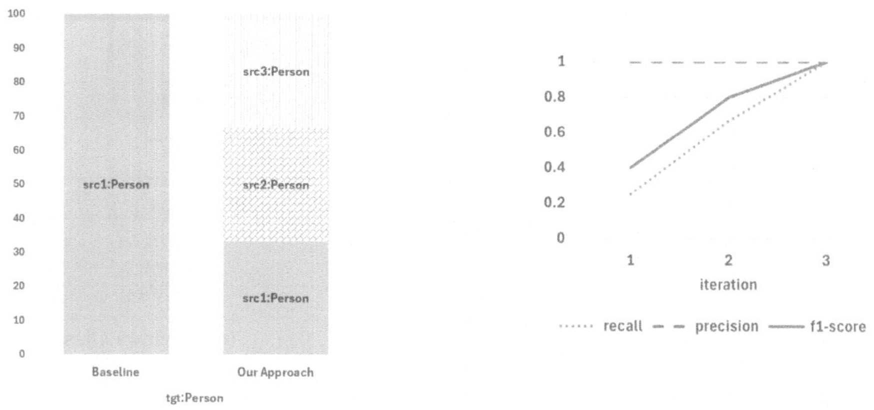

(a) Source Types of the Identified Candidate Instances

(b) Impact of the Number of Iterations on the Quality of the Result

Fig. 4. Candidate Instances for the Target Class Person in Scenario 2

5.4 Evaluation of the Impact of the Similarity Threshold

Figures 5 and 6 show the impact of the similarity threshold on the f1-score by comparing the result sets in scenarios 1 and 2, respectively.

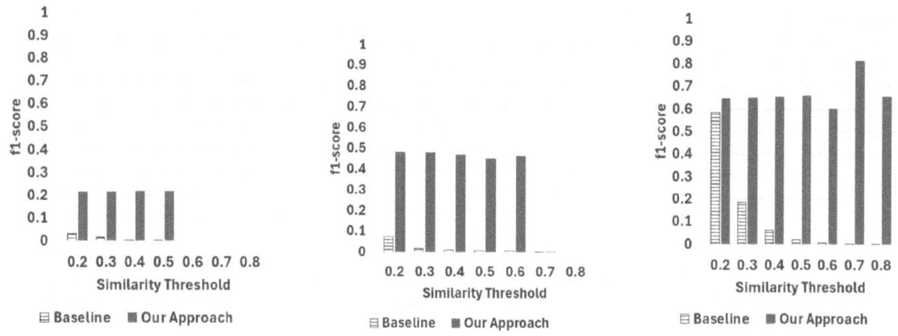

(a) Publisher Class.

(b) Book Class.

(c) Writer Class.

Fig. 5. Impact of the Similarity Threshold on the Target Schema Instantiation

Figure 5 shows the f1-score obtained with different similarity thresholds for each target class in scenario 1. In all cases (for each target class), our approach achieves a higher f1-score than the baseline.

Figure 6 shows that both our approach and the baseline achieve a good f1-score when the similarity threshold is lower than 0.4. However, when the similarity threshold is higher than 0.4, the f1-score for our approach is higher than the

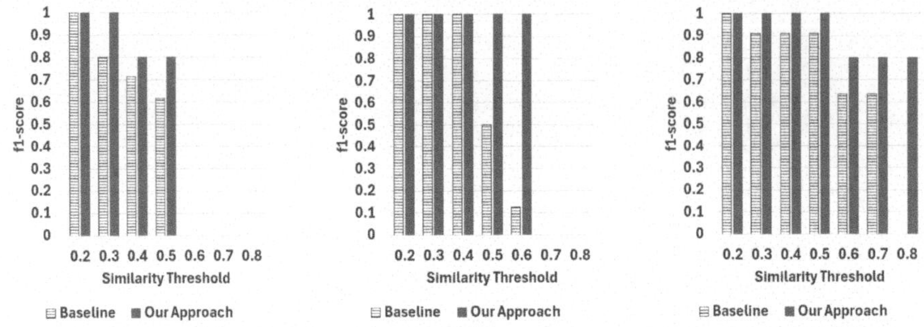

(a) Person Class. (b) Product Class. (c) Reviews Class.

Fig. 6. Impact of the Similarity Threshold on the Target Schema Instantiation

one of the baseline approach. The fact that the datasets are very homogeneous explains why the baseline achieves good results as well. These experiments show that our approach has better results for heterogeneous data sources.

5.5 Evaluation of the Quality Through Iterations

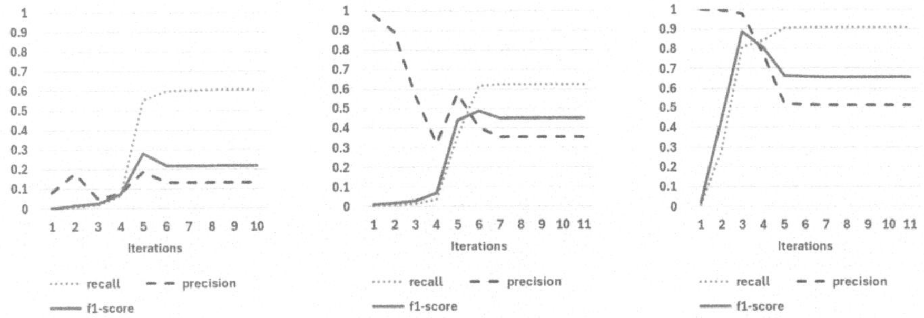

(a) Publisher Class. (b) Book Class. (c) Writer Class.

Fig. 7. Impact of the Number of Iterations on the Quality of the Results

Figure 7 depicts the evolution of the quality of the candidate instance descriptions through the different iterations by comparing the recall, precision, and f1-score at each iteration. The similarity threshold is set to 0.5. Figure 7 shows that the recall significantly improves during the first iterations. In Fig. 7b and 7c, the precision sharply decreases while it is quite stable in Fig. 7a. In all cases, after some iterations, the results stabilize. Considering these results, we could extend our approach by introducing a quality-based stopping criterion.

6 Related Works

The problems addressed by data discovery approaches [2,6,9] are somehow related to our work. The goal of these approaches is to identify in data lakes the relationships among datasets, such as joinable or unionable tables for a given target table. Unlike our work, these approaches usually deal with structured data (e.g., CSV files). They rely on a predefined schema for both the target and the source, which are not always provided in our setting.

One family of approaches that can be compared to our proposal is mapping generation, which has been introduced by Fagin et al. [4]. Recent approaches for mapping generation address problems such as missing integrity constraints [8]. Most of these approaches deal with structured data. The generation of mapping between ontologies or knowledge bases has been the subject of some research works [1,10,12]. Sacramento et al. [12] defined an approach that produces mappings between a set of local ontologies and a domain ontology. Some approaches [1,10] propose the generation of SPARQL executable mappings. These works assume that alignments between the source schemas and the target schema are provided, generated by some schema matching tools [3]. In our approach, we consider that schema alignments are not always provided. To the best of our knowledge, the generation of SPARQL executable mappings when the schema alignment is missing or incomplete has not been addressed.

7 Conclusion and Future Works

In this paper, we have proposed a novel approach to identify the relevant portion of RDF data sources considering some requirements expressed as a target schema, and to generate the SPARQL queries in order to extract these relevant data automatically. The proposed algorithms succeed in instantiating the target schema even when the alignment between the target schema and the source schema is not provided.

Our approach identifies the candidate instance descriptions by comparing the property sets of the source entities and the target classes using both schema and instance based similarity measures. The entities are iteratively compared to the identified candidate instance descriptions using a semi-supervised learning algorithm. The SPARQL query generation is performed on a partition of the resulting candidate instance descriptions.

The results of our experiments show that our approach outperforms a baseline approach which relies on schema-based similarity and exploits schema alignment.

In our future works, we will extend our approach by defining a stopping criterion for the iterative process by assessing the quality of the generated candidate instance descriptions at each iteration. In order to define a similarity threshold, we will study the correlations between the characteristics of the data sources and the target schema.

Moreover, we are currently extending our approach by considering some constraints defined on the target schema to extract only the candidate instance descriptions for which the constraints are verified. These constraints are defined using the SHACL[7] language.

Finally, we will also explore the possible ways of addressing the scalability of target schema instantiation, which may involve the design of parallel algorithms, or the use of indexation techniques.

References

1. Bashardoost, B.G., Miller, R.J., Lyons, K.A., Nargesian, F.: Knowledge translation: extended technical report. CoRR abs/2008.01208 (2020)
2. Bogatu, A., Fernandes, A.A.A., Paton, N.W., Konstantinou, N.: Dataset discovery in data lakes. In: 36th IEEE International Conference on Data Engineering, ICDE 2020, Dallas, TX, USA, 20–24 April 2020, pp. 709–720. IEEE (2020)
3. Euzenat, J., Shvaiko, P.: Ontology Matching, 2nd edn. Springer, Heidelberg (2013)
4. Fagin, R., Haas, L.M., Hernández, M., Miller, R.J., Popa, L., Velegrakis, Y.: Clio: schema mapping creation and data exchange. In: Borgida, A.T., Chaudhri, V.K., Giorgini, P., Yu, E.S. (eds.) Conceptual Modeling: Foundations and Applications. LNCS, vol. 5600, pp. 198–236. Springer, Heidelberg (2009). https://doi.org/10.1007/978-3-642-02463-4_12
5. Fernández, J.D., Martínez-Prieto, M.A., de la Fuente Redondo, P., Gutierrez, C.: Characterising RDF data sets. J. Inf. Sci. **44**(2), 203–229 (2018)
6. Fernandez, R.C., Abedjan, Z., Koko, F., Yuan, G., Madden, S., Stonebraker, M.: Aurum: a data discovery system. In: 34th IEEE International Conference on Data Engineering, ICDE 2018, Paris, France, 16–19 April 2018. pp. 1001–1012. IEEE Computer Society (2018)
7. Jaccard, P.: Distribution de la flore alpine dans le Bassin des Dranses et dans quelques régions voisines (1901)
8. Mazilu, L., Paton, N.W., Fernandes, A.A.A., Koehler, M.: Schema mapping generation in the wild. Inf. Syst. **104**, 101904 (2022)
9. Nargesian, F., Zhu, E., Pu, K.Q., Miller, R.J.: Table union search on open data. Proc. VLDB Endow. **11**(7), 813–825 (2018)
10. Rivero, C.R., Hernández, I., Ruiz, D., Corchuelo, R.: Generating SPARQL executable mappings to integrate ontologies. In: Jeusfeld, M., Delcambre, L., Ling, T.-W. (eds.) ER 2011. LNCS, vol. 6998, pp. 118–131. Springer, Heidelberg (2011). https://doi.org/10.1007/978-3-642-24606-7_10
11. Rivero, C.R., Schultz, A., Bizer, C., Ruiz, D.: Benchmarking the performance of linked data translation systems. In: Bizer, C., Heath, T., Berners-Lee, T., Hausenblas, M. (eds.) WWW2012 Workshop on Linked Data on the Web, Lyon, France, 16 April, 2012. CEUR Workshop Proceedings, vol. 937. CEUR-WS.org (2012)
12. Sacramento, E.R., Vidal, V.M.P., de Macêdo, J.A.F., Lóscio, B.F., Lopes, F.L.R., Casanova, M.A.: Towards automatic generation of application ontologies. J. Inf. Data Manag. **1**(3), 535–550 (2010)
13. Yarowsky, D.: Unsupervised word sense disambiguation rivaling supervised methods. In: Proceedings of the 33rd Annual Meeting on Association for Computational Linguistics, pp. 189–196. Association for Computational Linguistics, Cambridge (1995)

[7] https://www.w3.org/TR/shacl/.

Algorithms and Optimization

A Recursive Approach for Maximal (Δ, γ)-Clique Enumeration in Temporal Networks

Bithika Pal(✉)

Department of Computer Science and Engineering, Indian Institute of Technology Kharagpur, Kharagpur 721302, West Bengal, India
bithikapal@iitkgp.ac.in

Abstract. A set of objects and a binary relation among them is often represented by a graph or network. Most of the networks that we deal with in practice (e.g., social networks, human contact networks, financial transaction networks, etc.) are time-varying in nature, i.e., the relationship changes over time. Such networks are often modeled as Temporal Networks. In this paper, we study the problem of enumerating cohesive substructures present in a given temporal network. We call such substructure as (Δ, γ)-clique, which is a tuple of vertex subset and time interval, such that every pair of vertices in the vertex subset has at least γ edges in every Δ duration of the time interval. We propose a recursive solution approach to enumerate all the maximal (Δ, γ)-cliques. The proposed approach is divided into two parts. First, it initializes all the cliques of size 2 with maximum duration satisfying the (Δ, γ)-clique property, and recursively, adds vertices till it becomes maximal. The correctness of the proposed method has been established, and the complexity analysis has also been done. Several experiments are carried out using real-world temporal network datasets to highlight the efficiency of the proposed approach. The reported results show that the proposed solution approach is approximately six times faster and more space-efficient than the best existing method.

Keywords: Temporal Network · $(\Delta \cdot \gamma)$-Clique · Enumeration Algorithm · Δ-Degeneracy · Maximal Clique

1 Introduction

A large number of data that is found in today's world is of the following type: a set of objects and a binary relationship among themselves. Such data can be represented as a graph where the vertex set constitutes the set of objects and the set of binary relationships forms the edge set of the graph. Examples include *social network* where the vertex set is the set of users and the edge set is the set of friendship relations, *transportation network* where the vertex set is the set of cities and the edge set is the set of roads connecting the cities, *World Wide*

J. Tekli et al. (Eds.): ADBIS 2024, LNCS 14918, pp. 79–92, 2024.
https://doi.org/10.1007/978-3-031-70626-4_6

Web Network [5] where the vertex set is the set of websites and the edge set contains the set of all hyperlinks etc. Analysis of such networks forms the basis of many disciplines including *social network analysis* [29], *computational biology* [19], *computational epidemiology* [17], and many more.

Analysis of a network is done for different purposes and among many one of them is to analyze the network for its different structural patterns. Several structural patterns have been introduced and studied in the literature including Clique [27], Clan [26], Club [11], Core [16], Plex [7], Quasi Clique [23], and many more. As per the scope of the paper, we are interested in clique which is defined as a subset of the vertices such that between every pair of vertices there must exist an edge. Given a graph and a positive integer k, deciding whether the graph contains a clique of size k or not is an NP-hard problem. Still finding and enumeration of cliques in a graph has great practical relevance. In the case of a social network, a clique may represent a group of people who are very close friends. Similarly, in the case of a protein-protein interaction network, a clique may represent a group of proteins that can dock with each other to form new proteins. Hence, the enumeration of maximal cliques for a given graph brings many important sub-structures of a graph.

Due to many practical applications, the enumeration of maximal cliques present in a graph has gotten attention from the researchers of the database and data management community for the last three decades and a plethora of literature is available. To the best of my knowledge, Eralp A. Akkoyunlu [1] was the first to address the maximal clique enumeration problem. Later, Bron and Kerbosch [6] proposed a recursive solution to this problem. These two studies triggered a huge amount of research and the maximal clique enumeration problem has been studied in different setting such as in sparse graphs [10], in large networks [8,9,22] in map-reduce framework [14,28], in uncertain graphs, in parallel and distributed computing framework and many more.

Many of the networks that emerge in practice are time-varying in nature, i.e., the existing edges go off and new edges come up. Hence, the structure of the network changes over time. Such networks are called as *Temporal Networks*. There are two typical approaches to viewing such networks. One of them is the snap-shot graph model where a temporal network can be viewed as a series of static graphs, one for each time stamp. The other model is called a link stream, where the network is viewed as a list of triplets describing the interactions between two nodes at different timestamps. In recent times, the notion of clique has been generalized in the context of temporal graphs. The initial contribution came from Virad et al. [25] who introduced the notion of Δ-Clique which is defined as the vertex subset-time interval pair, where every pair of vertices must form an edge in every Δ duration of the time interval. They also showed that the notion of Δ-Clique can be helpful for analyzing the human contact network datasets and brings out interesting contact patterns among people. Later, the notion of Δ-Clique has been extended to (Δ, γ)-Clique by Banerjee and Pal [2,3] by introducing the frequency parameter γ. Their study showed that the notion of (Δ, γ)-Clique can be interpreted as a temporal community. Later several variants of clique have been introduced and studied in the literature such as periodic cliques [21], isolated cliques [20], interval cliques [12], atomic cliques [15]. There are several studies on the temporal clique enumeration Problem. Himmel et al. [13] adopted the Bron-Kerbosch algorithm for enumerating maximal cliques in temporal graphs. Zhu et al. [30] studied the temporal clique enumeration problem and proposed one solution approach. In this paper, we propose a (Δ, γ)-clique enumeration algorithm which adopts the famous Bron-Kerbosch Algorithm [6]. In this paper, we develop one such algorithm and in particular we make the following contributions in this paper:

- We propose an enumeration algorithm for maximal (Δ, γ)-Cliques for a given temporal network.
- We show that the proposed solution approach is correct.
- A number of experiments have been conducted with three real-world, publicly available temporal network datasets and the reported results show significant improvement both in terms of computational time and space from the existing state-of-the-art algorithms.

The rest of the paper is organized as follows. Section 2 describes required preliminary concepts and defines our problem formally. Section 3 describes the proposed solution approach with a detailed analysis. Section 4 describes the experimental evaluation of the proposed solution approach. Finally, Sect. 5 concludes our study.

2 Preliminaries and Problem Definition

In this section, we describe the preliminary concepts and define our problem formally. All the graphs considered in this paper are simple, finite, and undirected. For any graph $G(V, E)$, $V(G)$, and $E(G)$ denotes its vertex and edge set, respectively. We denote the number of vertices and edges of G by n and m, respectively, i.e., $|V(G)| = n$ and $|E(G)| = m$. For any vertex $v \in V(G)$, $N(v)$ denotes the set of other vertices which are adjacent to v, i.e., $N(v) = \{u : (uv) \in E(G)\}$. The degree of a vertex is defined as the cardinality of its neighborhood, i.e., $deg(v) = |N(v)|$. Now, we state the notion of clique in a graph in Definition 1.

Definition 1 (Clique). *Given a graph $G(V, E)$, a subset of vertex $S \subseteq V(G)$ is said to be a clique in a graph if every pair of vertices in S forms an edge in the graph G.*

Given a graph $G(V, E)$ and positive integer k, it is an NP-hard problem to decide whether G contains a clique of size k or not. A clique is said to be maximal if no more vertices can be added to it. Among the maximal cliques, one with the maximum size is called the maximum size clique.

As mentioned previously, most of the networks that we deal with in practice, are time-varying in nature. We define the notion of the time-varying network in Definition 2.

Definition 2 (Temporal Network). *A temporal network is defined as $G(V, E, \mathcal{T})$, where $V(G)$ denotes the set of vertices, $E(G)$ denotes the set of temporal edges where each entry is a triplet of the form (t, u, v) signifying the connection between vertex u and v at timestamp t ($t \in \mathcal{T}$), and $\mathcal{T}$ denotes the set of discrete time stamps in which the network is observed.*

In this study, we consider that the set of discrete time stamps in which the network is observed is: $\mathcal{T} = \{t_{min}, \dots, t-1, t, t+1, \dots, t_{max}\}$. A temporal network can be represented in many ways such as link stream, series of snapshot graphs, and weighted graph. In this work, we consider the weighted one, where the weights W, store the set of occurrence timestamps (in increasing order) of a static edge in the de-temporal graph. Mathematically, $W(G) : V(G) \times V(G) \longrightarrow \{t | (t, u, v) \in E(G) \wedge u, v \in V(G)\}$

The notion of clique has been generalized in the context of temporal graphs. Virad et al. [25] introduced the notion of Δ-Clique which has been stated in Definition 3.

Definition 3 (Δ-Clique). *Given a temporal network $G(V, E, \mathcal{T})$, a vertex subset-time interval pair $(X, [t_a, t_b])$ is said to be a Δ-Clique if between every pair of vertices in X there must exist one edge in every Δ duration of the time interval $[t_a, t_b]$.*

Later, the notion of Δ-Clique has been extended by adding a frequency parameter γ by Banerjee and Pal [2] which has been stated in Definition 4.

Definition 4 ((Δ, γ)-Clique). *Given a temporal network $G(V, E, \mathcal{T})$, a vertex subset-time interval pair $(X, [t_a, t_b])$ is said to be a (Δ, γ)-Clique if between every pair of vertices in X there must exist at least γ edges in every Δ duration of the time interval $[t_a, t_b]$.*

There are two conditions for maximality of a (Δ, γ)-Clique and this has been described in Definition 5.

Definition 5 (Maximal (Δ, γ)-Clique). *Given a temporal network $G(V, E, \mathcal{T})$, a (Δ, γ)-Clique $(X, [t_a, t_b])$ is said to be maximal if neither of the following three statements is true:*

- $\exists v \in V(G) \setminus X$, $(X \cup \{v\}, [t_a, t_b])$ *is a (Δ, γ)-Clique.*
- $(X, [t_a - 1, t_b])$ *is a (Δ, γ)-Clique. This condition is applied only when $t_a - 1 \geq 1$.*
- $(X, [t_a, t_b + 1])$ *is a (Δ, γ)-Clique. This condition is applied only when $t_b + 1 \leq T$.*

The clique is said to be *duration-wise maximal* if it satisfies all the conditions except the first one. The (Δ, γ)-Cliques present in a temporal network can be viewed as the communities. Hence, the problem of enumeration of Maximal (Δ, γ)-Cliques present in a temporal network is of immense importance and this problem has been stated in Definition 6.

Definition 6 (Maximal (Δ, γ)-Clique Enumeration Problem). *Given a temporal network $G(V, E, \mathcal{T})$, the problem of Maximal (Δ, γ)-Clique Enumeration asks to list out all the maximal (Δ, γ)-cliques present in the given temporal network G.*

We introduce the following definitions which are needed to understand the recursive solution approach of the problem.

Definition 7 ((Δ, γ)-Maximal Duration Edge-Weight). *For a given set of temporal occurrences of a static edge, $W(u, v)$, the (Δ, γ)-Maximal Duration edge-weight, $W^{(\Delta,\gamma)}(u, v)$, is the set of intervals $[t_a, t_b]$ with the property that for every $t \in [t_a, t_b - \Delta]$ at least γ links corresponding to the static edge $\{u, v\}$ exist in each Δ duration starting from t. The duration $[t_a, t_b]$ is maximal, (i.e., neither $[t_a - 1, t_b]$ nor $[t_a, t_b + 1]$ has γ links in each of its Δ duration), and $t_b - t_a \geq \Delta$.*

Definition 8 ((Δ, γ)-Neighborhood). *For a vertex $v \in V$ and a time interval $[t_a, t_b] \subseteq \mathcal{T}$, the (Δ, γ) neighborhood, $N^{(\Delta,\gamma)}(v, [t_a, t_b])$, is the set of all vertex interval pairs $(u, [t_a^{'}, t_b^{'}])$ with the property that for every $t \in [t_a^{'}, t_b^{'} - \Delta]$ at least γ links corresponding to the static edge $\{u, v\}$ exist in each Δ duration starting from t. The duration $[t_a^{'}, t_b^{'}]$ is maximal, (i.e., there does not exist any $[t_a^{''}, t_b^{''}] \subseteq [t_a, t_b]$ with $[t_a^{''}, t_b^{''}] \supset [t_a^{'}, t_b^{'}]$), and $t_b^{'} - t_a^{'} \geq \Delta$.*

We can say, $N^{(\Delta,\gamma)}(v, [t_a, t_b]) = \{(u, [t_a^{'}, t_b^{'}]) | u \in N(v) \wedge [t_a^{'}, t_b^{'}] \in \{[t_a, t_b]\} \cap_{\Delta} W^{(\Delta,\gamma)}(u, v)\}$, where $N(v)$ denotes the neighbors of v in the de-temporal graph extracted from the temporal network G. The Δ-intersect between two sets of intervals A and B, $A \cap_{\Delta} B$, computes the element-wise intersections of the intervals and keeps the common interval having length at least Δ.

Definition 9 (Temporal Membership). *For a given set of vertex-interval pairs, $\mathcal{A}$, the temporal membership of an entry $(v, [t_a, t_b])$ checks the existence of a member $(v, [t_a^{'}, t_b^{'}]) \in \mathcal{A}$ such that $[t_a, t_b] \subseteq [t_a^{'}, t_b^{'}]$. We denote this as $(v, [t_a, t_b]) \in_{temp} \mathcal{A}$.*

Definition 10 (Δ-Cut). *For two given sets of vertex-interval pairs, $\mathcal{A}$ and $\mathcal{B}$, the Δ-Cut between them have all the vertex-interval pairs, where the vertex is present in both the sets (in an entry) and the interval is the common interval between two entries having length at least Δ. We denote this as $\mathcal{A} \cap_{cut} \mathcal{B}$.*

Mathematically, $\mathcal{A} \cap_{cut} \mathcal{B} = \{(v, [t_a^{'}, t_a^{''}] \cap [t_b^{'}, t_b^{''}]) | (v, [t_a^{'}, t_a^{''}]) \in \mathcal{A} \wedge (v, [t_b^{'}, t_b^{''}]) \in \mathcal{B} \wedge |[t_a^{'}, t_a^{''}] \cap [t_b^{'}, t_b^{''}])| \geq \Delta\}$. We can re-write this using temporal membership, $\mathcal{A} \cap_{cut} \mathcal{B}$ contains all the vertex interval pairs $(v, [t^{'}, t^{''}])$ of maximal duration where $(v, [t^{'}, t^{''}]) \in_{temp} \mathcal{A}$, $(v, [t^{'}, t^{''}]) \in_{temp} \mathcal{B}$,and $t^{''} - t^{'} \geq \Delta$.

3 Proposed Solution Approach

In this section, we describe the proposed solution approach for the maximal (Δ, γ)-clique enumeration problem. The proposed solution approach takes a temporal network, and Δ, γ values, and returns the set of maximal (Δ, γ)-cliques contained in it. The approach is divided into two parts; (i) Computation of (Δ, γ)-maximal duration edge weight $W^{(\Delta,\gamma)}$ for each static edge in the graph, (ii) maximal clique formation using $W^{(\Delta,\gamma)}$. As per Definition 7, $W^{(\Delta,\gamma)}$ is nothing but the duration-wise maximal 2-size cliques. So, we use Algorithm 1 from [4] to compute such clique set $\mathcal{C}_T^I$ in Line 1. Now, propagate the dictionary $W^{(\Delta,\gamma)}$ from Line 2–5 in Algorithm 1.

Algorithm 1: Enumeration of all Maximal (Δ, γ)-Cliques in a Temporal Network

Input: Temporal network $G(V, E, \mathcal{T})$, Δ, and γ values
Output: Set of Maximal (Δ, γ)-Cliques, $\mathcal{C}$

1 $\mathcal{C}_T^I \longleftarrow$ Set of 2 size duration-wise maximal $(\Delta, \gamma) - Cliques$;
2 $W^{(\Delta,\gamma)} \leftarrow$ empty dictionary;
3 **for** *each* $(\{u, v\}, [t_a, t_b]) \in \mathcal{C}_T^I$ **do**
4   Add $[t_a, t_b]$ into $W^{(\Delta,\gamma)}(u, v)$;
5 **end**
6 $\mathcal{P} \leftarrow \emptyset$, $\mathcal{S} \leftarrow \emptyset$, $\mathcal{C} \leftarrow \emptyset$; `//` $\mathcal{C}$ `and` $W^{(\Delta,\gamma)}$ `are global structures`
7 $\tilde{t}_{min} \leftarrow t_{min} - \Delta + (\gamma - 1), \tilde{t}_{max} \leftarrow t_{max} + \Delta - (\gamma - 1)$;
8 $\forall v \in V(G)$: add $(v, [\tilde{t}_{min}, \tilde{t}_{max}]) \in \mathcal{P}$;
9 $C_I \leftarrow (\emptyset, [\tilde{t}_{min}, \tilde{t}_{max}])$;
10 `BronKerboschDeltaGamma(`C_I, $\mathcal{P}$, $\mathcal{S}$`)`;
11 **Return** $\mathcal{C}$;

12 **function** `BronKerboschDeltaGamma(`C, $\mathcal{P}$, $\mathcal{S}$`)`:
    `//` $C = (X, [t_a, t_b])$ `: duration-wise maximal` (Δ, γ)`-clique`
    `//` $\mathcal{P} \cup \mathcal{S}$`: set of all` $(v, [t_a^{'}, t_b^{''}])$ `s.t.` $[t_a^{'}, t_b^{'}] \subseteq [t_a, t_b]$ `and` $(X \cup \{v\}, [t_a^{'}, t_b^{'}])$ `is a duration-wise maximal` (Δ, γ)`-clique`
13   **if** $\forall (v, [t_a^{'}, t_b^{'}]) \in \mathcal{P} \cup \mathcal{S} : [t_a^{'}, t_b^{'}] \subsetneq [t_a, t_b]$ **then**
14   add $(X, [t_a, t_b])$ to the solution $\mathcal{C}$;
15   **end**
    `// choose pivot element`
16     $(v_p, [t_p^{'}, t_p^{''}]) \leftarrow \underset{(v, [t_a^{'}, t_b^{'}]) \in \mathcal{P} \cup \mathcal{S}}{argmax} |\mathcal{P} \cap_{cut} N^{(\Delta,\gamma)}(v, [t_a^{'}, t_b^{'}])|$;
17   **for** *each* $(v, [t_a^{'}, t_b^{'}]) \in \mathcal{P} \setminus \{(u, [t_a^{''}, t_b^{''}]) : (u, [t_a^{''}, t_b^{''}]) \in \mathcal{P} \wedge (u, [t_a^{''}, t_b^{''}]) \in_{temp} N^{(\Delta,\gamma)}(v_p, [t_p^{'}, t_p^{''}])\}$ **do**
18         $C^{'} \leftarrow (X \cup \{v\}, [t_a^{'}, t_b^{'}])$;
19         $\mathcal{P}^{'} \leftarrow \mathcal{P} \cap_{cut} N^{(\Delta,\gamma)}(v, [t_a^{'}, t_b^{'}])$;
20         $\mathcal{S}^{'} \leftarrow \mathcal{S} \cap_{cut} N^{(\Delta,\gamma)}(v, [t_a^{'}, t_b^{'}])$;
21   `BronKerboschDeltaGamma(`$C^{'}$, $\mathcal{P}^{'}$, $\mathcal{S}^{'}$`)`;
22         $\mathcal{P} \leftarrow \mathcal{P} \setminus (v, [t_a^{'}, t_b^{'}])$;
23         $\mathcal{S} \leftarrow \mathcal{S} \cup (v, [t_a^{'}, t_b^{'}])$;
24   **end**

The function `BronKerboschDeltaGamma()` in Algorithm 1 computes the maximal (Δ, γ)-Clique by maintaining currently computed clique $C = (X, [t_a, t_b])$, and two sets of vertex-interval pairs $\mathcal{P}$ and $\mathcal{S}$. $\mathcal{P}$ contains potential candidates which can be used to extend $(X, [t_a, t_b])$, and $\mathcal{S}$ contains the candidates that are excluded from consideration in the current clique being constructed. If for all the vertex-interval pair in $\mathcal{P} \cup \mathcal{S}$, the interval exactly lies within $[t_a, t_b]$, then any of the vertex addition will generate a duration less than $[t_a, t_b]$. So, $[t_a, t_b]$ becomes the maximal duration for X. Hence, $(X, [t_a, t_b])$ is the maximal one and added to the solution set $\mathcal{C}$ [Refer Line 13–15]. Next, we choose a pivot element from $\mathcal{P} \cup \mathcal{S}$ in Line 16 to minimize the number of recursive calls for the current clique $(X, [t_a, t_b])$ [Line 16]. For each potential vertex-interval pair $(v, [t_a^{'}, t_b^{'}])$ in $\mathcal{P}$, the current clique $(X, [t_a, t_b])$ is considered for vertex expansion within the duration $[t_a^{'}, t_b^{'}]$ from Line 17 to 24. To check this, new potential set of candidates $\mathcal{P}^{'}$ and set of excluded candidates $\mathcal{S}^{'}$ are formed using $\mathcal{P} \cap_{cut} N^{(\Delta,\gamma)}(v, [t_a^{'}, t_b^{'}])$ and $\mathcal{S} \cap_{cut} N^{(\Delta,\gamma)}(v, [t_a^{'}, t_b^{'}])$, respectively [Line 19–20]. As discussed in Definition 8, $N^{(\Delta,\gamma)}(v, [t_a^{'}, t_b^{'}])$ can be computed from $W^{(\Delta,\gamma)}$. The function `BronKerboschDeltaGamma()` is called in Line 21 for the new clique $(X \cup \{v\}, [t_a^{'}, t_b^{'}])$, with the corresponding potential and excluded set of candidates $\mathcal{P}^{'}$ and $\mathcal{S}^{'}$. The recursion continues till the base case arrives. Once all the vertex-interval pairs are checked in depth, it back-tracks to the original function and removes the considered $(v, [t_a^{'}, t_b^{'}])$ from $\mathcal{P}$ and adds it to the excluded set of candidates $\mathcal{S}$. We start the recursion with an initial clique (C_I) having an empty vertex set and the entire time horizon as its duration. The set of potential vertex-interval pairs $\mathcal{P}$ for C_I is all the vertices in G associated with the entire time horizon. As there are no candidates to be excluded, $\mathcal{S}$ is initialized empty set. Algorithm 1 calls `BronKerboschDeltaGamma()` with this initialization and returns the final solution set $\mathcal{C}$ containing all the maximal (Δ, γ)-Cliques present in the given temporal network.

Correctness of the Proposed Algorithm. As we extend the recursive approach from [13], the correctness of the $N^{(\Delta,\gamma)}$ formation gives the correctness of the proposed algorithm. Also, we need to show that the Δ-cut operator returns the vertex interval pair, where each vertex has appeared at least γ times in each Δ duration of the interval, this is easy to follow as it computes the Δ-intersection of each interval. Hence, if any of the participating sets in the Δ-cut operator follows the (Δ, γ) condition, the resulting set will also follow the same. Now, we only need to show that $N^{(\Delta,\gamma)}$ follows the (Δ, γ) condition. Now, $N^{(\Delta,\gamma)}$ is populated using $W^{(\Delta,\gamma)}$. The $N^{(\Delta,\gamma)}(v, [t_a, t_b])$ is the set $\{(u, [t_a^{'}, t_b^{'}]) | u \in N(v) \wedge [t_a^{'}, t_b^{'}] \in \{[t_a, t_b]\} \cap_\Delta W^{(\Delta,\gamma)}(u, v)\}$. Now $W^{(\Delta,\gamma)}$ computes correctly all the maximal duration for each static edge [refer Lemma 6 [4]], $N^{(\Delta,\gamma)}$ will also be correct. Hence, the proposed methodology correctly enumerates all the maximal (Δ, γ)-Cliques in the given temporal network.

Complexity Analysis. It has been reported in the literature that a graph with *degeneracy* d can have at most $\mathcal{O}(n \cdot 3^{\frac{d}{3}})$ many maximal cliques and they can be enumerated in $\mathcal{O}(d \cdot n \cdot 3^{\frac{d}{3}})$ [10]. To initialize the size 2 cliques at Line 1 of

Algorithm 1 will take $\mathcal{O}(\gamma \cdot m)$ time. As the cardinality of $\mathcal{C}_I^T$ can be of $\mathcal{O}(m \cdot |\mathcal{T}|)$, hence the `for` loop at Line 3 to 5 will execute for $\mathcal{O}(m \cdot |\mathcal{T}|)$ times. Instructions in Line 6 and 7 will take $\mathcal{O}(1)$ time and Line 8 will take $\mathcal{O}(n)$ time. At Line 10, the recursive function for clique computation has been invoked and the running time of this function is analyzed with the help of Δ-slice degeneracy.

The Δ-slice degeneracy of a temporal graph will be d if $\forall t \in \mathcal{T}$, every snapshot graph $G_t(V, E_t)$, where $E_t = \{(t', u, v) | t' \in [t, t+\Delta] \wedge u, v \in V(G_t)\}$, has at most degeneracy d. In each recursive call, we pass a duration-wise maximal clique and for converting a duration-wise maximal to maximal clique one extra function call is required. Hence, the total running time of the function becomes $\mathcal{O}(x \cdot m)$, where x signifies the number of duration-wise maximal cliques. Now, in each G_t there can be at most $n \cdot 3^{\frac{d}{3}}$ maximal static clique and $2^{d+1} \cdot n \cdot 3^{\frac{d}{3}}$ all sized cliques. For each static clique, there can be at most $|\mathcal{T}|$ duration-wise maximal cliques. Hence, x is bounded by size of $\mathcal{O}(2^{d+1} \cdot n \cdot 3^{\frac{d}{3}} \cdot |\mathcal{T}|)$ The $N^{(\Delta,\gamma)}$ can be computed in $\mathcal{O}(\gamma \cdot m)$ time. Also, for any two sets $\mathcal{A} \cap_{cut} \mathcal{B}$ can be computed in $\mathbf{O}(|\mathcal{A}| + |\mathcal{B}|)$ [13]. The pivot is calculated in $\mathcal{O}(n^2 \cdot |\mathcal{T}|^2)$. Hence by keeping the dominating term we can conclude the time complexity of Algorithm 1 is $\mathcal{O}(2^{d+1} \cdot n \cdot 3^{\frac{d}{3}} \cdot |\mathcal{T}| \cdot m + n^2 \cdot |\mathcal{T}|^2)$.

4 Experimental Evaluation

Dataset Description. In this study, we have used the following datasets:

- **High School Dataset**[1] [18]: This dataset gives the interactions between 327 students of nine classes within a high school in Marseilles, France, in December 2013, during five days in December 2013.
- **Primary School Dataset**[2] [24]: The dataset is collected in a French primary school on October 1 and 2, 2009 using RFID-based instruments. It contains the contact patterns among 232 children and 10 teachers.
- **Haggle**[3]: The network represents contacts between people carrying wireless devices. A node represents a person; an edge between two persons shows that there was a contact between the devices.

Table 1. Dataset Description

Dataset	#Nodes	#Edges	#Links	Lifespan
High School	327	5818	188508	5 days
Primary School	242	8317	125773	2 days
Haggle	274	2899	28244	4 days

[1] http://www.sociopatterns.org/datasets/high-school-contact-and-friendship-networks/.
[2] http://www.sociopatterns.org/datasets/primary-school-temporal-network-data/.
[3] http://konect.cc/networks/contact/.

Table 1 contains the basic statistics of the datasets. These datasets have been used by many previous studies on temporal network analysis (Fig. 1).

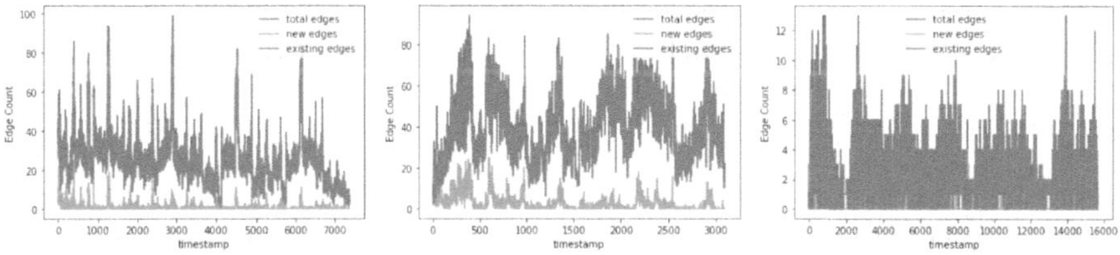

(a) High School (b) Primary School (c) Haggle

Fig. 1. Timestamp vs. edge arrival plots for different datasets

Goals of the Experiments. In our experiments, we fix the following research questions:

- RQ1: Once we fix a Δ value is fixed and γ value is increased then how the number of maximal (Δ, γ)-cliques, maximum duration, maximum cardinality are changing?
- RQ2: Once we fix the γ value and the Δ value is increased then how the number of maximal (Δ, γ)-cliques, maximum duration, maximum cardinality are changing?
- RQ3: How do the computational resource requirements (i.e., time and space) vary across proposed as well as comparing methods?

Comparing Algorithms. We have compared the performance of the proposed approach with the following existing solution approaches:

- Snapshot: In this method, the first snapshot graphs have been constructed at each time stamp t using the links from $[t, t+\Delta]$. Now, in each snapshot graph, we compute all the cliques. Now these cliques are merged along the time horizon to obtain the duration-wise maximal cliques. Finally, these duration-wise maximal cliques have been converted to the maximal (Δ, γ)-cliques.
- Iterative [25]: This method has been proposed by Viard et al. [25]. This solution approach first generates 2 size Δ duration cliques and expands them by both vertex addition and time interval stretching.
- Two Phase [4]: This method has been proposed by Banerjee and Pal [4]. In this method, first size 2 duration-wise maximal cliques are formed and subsequently, these cliques have been made maximal via vertex addition.

All the methods have been implemented in Python 3.7 with NetworkX 2.3 environment. All the experiments have been conducted in a server running on a Linux Environment with 128 GB of RAM and Intel(R) Xeon(R) CPU E5-2697 v2 @

2.70 GHz processor. The code is available at https://github.com/BITHIKA1992/Recursive_Delta_gamma_clique.

Experimental Results and Discussions. Here we describe the results obtained from our experiments. Table 2 records the clique count, clique size (maximum and average), and clique duration (maximum and average) for different Δ and γ values for different datasets. From this table, we make the following observations:

1. For any dataset, for a fixed Δ if the value of γ is increased then the clique count, clique size, and clique duration are decreased. As an example, for the high school dataset, for $\Delta = 60$, when γ is increased from 2 to 12, the clique count value is reduced from 13316 to 3457, the maximum clique count value has been decreased from 7 to 5, and the maximum clique duration has been reduced from 1786 to 1328. Similar observations have been made for other datasets as well. The reason behind this is quite intuitive.
2. For any dataset, for a fixed γ, once the Δ value has been increased, the clique count, clique size, and clique duration are increased. As an example for the Primary School dataset, when the γ value is fixed at 2 and the Δ value has been increased from 90 to 540, the number of maximal cliques has been increased from 40162 to 107801, maximum clique size has been increased from 10 to 17, and the maximum clique duration has been increased from 1469 to 2622.

The above description answers the research questions 1 and 2. Now, we proceed to describe the research question 3. Table 3 records the computational time and space requirement for different datasets and Δ, γ settings. Here, OOT and OOM signify the out-of-time and out-of-memory. After executing the code for 72 h, if we do not receive the results then we report that entry as OOT. In that case, the corresponding memory requirement will not be available and we report this by writing NA. Similarly, in many instances, the code abruptly terminates by reporting the out-of-memory error. In this case, the corresponding time requirement will not be available and we report this by writing NA.

In Table 3, from the instances where the results are available, it can be observed that the proposed solution approach takes much less computational time and space compared to the existing methods. As an example, for the High School dataset, for $\Delta = 2160$ and $\gamma = 18$, among the existing methods, the two-phase approach takes 14.19 secs whereas the proposed solution approach takes 2.29 secs. Hence, in this instance, the proposed solution approach is approximately six times faster than the best available solution approach. In this instance the space requirements by the two-phase and the proposed approach 124.62 and 23.45 (which is approximately one-fifth of the space requirement by the two-phase method), respectively.

Table 2. No. of Cliques, Maximum Duration, Maximum Cardinality of the (Δ, γ)-Clique Set for different dtasets

			Clique	Clique size		Clique Duration	
Data	Δ	γ	**Count**	**Maximum**	**Average**	**Maximum**	**Average**
HighSchool	90	2	13316	7	2.48	1786	209.35
	90	6	5735	5	2.22	1776	215.58
	90	12	3457	5	2.12	1328	209.71
	90	18	2589	4	2.09	1003	198.74
	540	2	14026	8	2.91	2695	1074.23
	540	6	4930	6	2.43	2687	1147.25
	540	12	2803	5	2.25	2675	1166.96
	540	18	2037	5	2.20	2663	1172.87
	2160	2	15883	8	3.19	5935	4214.57
	2160	6	5112	6	2.55	5927	4356.92
	2160	12	2861	5	2.32	5915	4395.37
	2160	18	2092	5	2.25	5903	4389.71
PrimarySchool	90	2	40162	10	3.45	1469	160.82
	90	6	7091	6	2.47	1458	175.20
	90	12	2851	6	2.26	734	168.36
	90	18	1630	5	2.20	603	161.30
	540	2	107801	17	5.82	2622	899.39
	540	6	14820	11	3.81	2604	926.77
	540	12	4324	7	2.77	2590	981.47
	540	18	2350	6	2.47	2575	1000.96
	2160	2	128841	20	7.19	5867	4008.62
	2160	6	24730	14	5.29	5857	3873.67
	2160	12	6535	10	3.62	5830	3934.51
	2160	18	2957	7	2.88	5815	4081.19
Haggle	1800	2	40752	23	7.31	23032	3192.70
	1800	4	15158	16	6.43	10967	2564.46
	1800	6	3962	8	3.93	5824	2219.50
	1800	8	701	5	2.58	4198	2077.08
	7200	2	72017	25	7.51	85633	12036.70
	7200	4	19575	20	6.52	46440	12510.77
	7200	6	7685	16	5.38	44963	12108.85
	7200	8	3230	12	4.07	41770	11733.60
	10800	2	102080	25	8.42	102646	17288.88
	10800	4	27064	20	6.59	69670	17746.67
	10800	6	9804	16	5.35	62401	17858.49
	10800	8	4065	12	4.11	49941	17626.12

Table 3. Results on comparing computational time and space for different (Δ, γ)-Clique Enumeration algorithms

			Computational Time (Sec.)				Computational Space (MB)			
Data	Δ	γ	Snapshot	Iterative	TwoPhase	Proposed	Snapshot	Iterative	TwoPhase	Proposed
HighSchool	90	2	1155.95	22372.16	56.66	10.41	370.582	5935.74	146.51	30.15
	90	6	507.98	3468.77	17.06	4.39	260.9844	3490.95	131.13	26.05
	90	12	1151.01	2016.53	8.90	2.99	235.9961	2655.39	127.50	24.29
	90	18	514.25	4912.98	7.24	2.51	228.5117	2116.47	123.88	23.79
	540	2	2980.77	4948.36	50.06	8.47	688.1406	11663.16	162.75	30.24
	540	6	778.59	20635.97	25.28	3.54	496.1328	4442.22	131.20	25.52
	540	12	832.73	4076.99	11.45	2.55	460.7656	3048.79	127.66	23.88
	540	18	587.14	3834.03	3.40	2.14	454.0547	2520.82	126.16	23.45
	2160	2	3184.46	43477.45	25.49	9.09	1584.652	21982.11	176.58	31.28
	2160	6	1804.04	11637.48	6.48	3.56	1223.914	5474.27	137.71	25.58
	2160	12	854.65	5300.35	11.36	2.45	1316.363	3273.93	127.71	23.88
	2160	18	764.14	8387.73	14.19	2.29	1308.238	2578.01	124.62	23.45
PrimarySchool	90	2	2145.54	11720.79	65.91	28.67	588.78	15448.15	286.04	44.82
	90	6	357.66	2130.23	9.82	4.91	296.07	3597.64	139.66	25.55
	90	12	133.00	750.55	4.02	2.24	283.50	1801.47	125.41	22.76
	90	18	83.40	392.28	2.29	1.59	280.79	1186.53	121.87	22.07
	540	2	83601.36	OOT	1199.34	126.08	2998.84	NA	3567.86	97.07
	540	6	1798.64	36546.30	40.00	8.55	812.37	26214.24	235.39	29.86
	540	12	495.71	4497.51	6.08	2.50	562.83	4654.11	135.19	23.47
	540	18	320.85	1936.03	6.05	1.72	546.51	2650.23	126.88	22.44
	2160	2	OOT	OOT	12333.22	437.86	NA	NA	28158.69	121.07
	2160	6	19942.74	OOT	185.39	23.63	3774.16	NA	846.70	39.50
	2160	12	2118.93	14938.79	16.19	3.68	1443.38	17892.49	166.58	25.31
	2160	18	963.58	2498.54	5.11	1.92	1340.26	4983.37	132.88	22.76
Haggle	1800	2	NA	NA	NA	268.76	OOM	OOM	OOM	44.88
	1800	4	NA	NA	1232.83	30.88	OOM	OOM	2257.52	26.09
	1800	6	2342.83	180.09	2.42	2.04	3807.37	1230.15	111.12	17.80
	1800	8	173.97	2.90	0.20	0.69	1607.18	120.67	87.63	14.83
	7200	2	NA	OOT	NA	986.08	OOM	NA	OOM	68.09
	7200	4	NA	OOT	709.07	73.60	OOM	NA	19072.80	29.58
	7200	6	NA	OOT	39.05	7.19	OOM	NA	1156.28	19.82
	7200	8	6170.54	3379.99	2.31	1.94	8960.05	10991.50	147.54	15.25
	10800	2	NA	OOT	OOT	524.95	OOM	NA	NA	96.73
	10800	4	NA	OOT	718.81	69.96	OOM	NA	19382.10	35.98
	10800	6	NA	OOT	44.10	9.78	OOM	NA	1295.04	20.82
	10800	8	7308.66	3973.54	2.65	2.02	10468.50	12423.08	158.30	18.18

5 Concluding Remarks

In this paper, we have studied the (Δ, γ)-Clique Enumeration Problem and proposed one solution approach that works well on temporal networks where the link to static edge ratio is high. The proposed methodology has been described in detail and an analysis of the time and space requirement has been done. A large number of experiments have been conducted with real-world, publicly available temporal network datasets and it can be observed that the proposed solution approach takes much less computational resources compared to the existing methods. Now, this work can be extended in several directions. The edges in the temporal graph may be probabilistic in nature. Hence, the maximal (Δ, γ)-clique enumeration problem can be modeled considering this phenomenon. Also, the notion of (Δ, γ)-clique can be applied in different contexts to bring out important contact patterns.

References

1. Akkoyunlu, E.A.: The enumeration of maximal cliques of large graphs. SIAM J. Comput. **2**(1), 1–6 (1973). https://doi.org/10.1137/0202001
2. Banerjee, S., Pal, B.: On the enumeration of maximal (Δ, γ)-cliques of a temporal network. In: Krishnapuram, R., Singla, P. (eds.) Proceedings of the ACM India Joint International Conference on Data Science and Management of Data, COMAD/CODS 2019, Kolkata, India, 3–5 January 2019, pp. 112–120. ACM (2019). https://doi.org/10.1145/3297001.3297015
3. Banerjee, S., Pal, B.: A two-phase approach for enumeration of maximal (Δ, γ)-cliques of a temporal network. In: Strauss, C., Kotsis, G., Tjoa, A.M., Khalil, I. (eds.) DEXA 2021. LNCS, vol. 12924, pp. 346–357. Springer, Cham (2021). https://doi.org/10.1007/978-3-030-86475-0_33
4. Banerjee, S., Pal, B.: A two-phase approach for enumeration of maximal (Δ, γ)-cliques of a temporal network. Soc. Netw. Anal. Min. **14**(1), 54 (2023). https://doi.org/10.1007/S13278-024-01207-Y
5. Barnett, G.A., Jiang, K.: Resilience of the world wide web: a longitudinal two-mode network analysis. Soc. Netw. Anal. Min. **6**, 1–15 (2016)
6. Bron, C., Kerbosch, J.: Algorithm 457: finding all cliques of an undirected graph. Commun. ACM **16**(9), 575–577 (1973)
7. Chang, L., Xu, M., Strash, D.: Efficient maximum k-plex computation over large sparse graphs. Proc. VLDB Endow. **16**(2), 127–139 (2022)
8. Cheng, J., Ke, Y., Fu, A.W.C., Yu, J.X., Zhu, L.: Finding maximal cliques in massive networks by h*-graph. In: Proceedings of the 2010 ACM SIGMOD International Conference on Management of Data, pp. 447–458 (2010)
9. Cheng, J., Ke, Y., Fu, A.W.C., Yu, J.X., Zhu, L.: Finding maximal cliques in massive networks. ACM Trans. Database Syst. (TODS) **36**(4), 1–34 (2011)
10. Eppstein, D., Löffler, M., Strash, D.: Listing all maximal cliques in large sparse real-world graphs. J. Experimental Algorithmics (JEA) **18**, 3–1 (2013)
11. Golovach, P.A., Heggernes, P., Kratsch, D., Rafiey, A.: Finding clubs in graph classes. Discret. Appl. Math. **174**, 57–65 (2014)
12. Hermelin, D., Itzhaki, Y., Molter, H., Niedermeier, R.: Temporal interval cliques and independent sets. Theor. Comput. Sci. **961**, 113885 (2023)
13. Himmel, A.S., Molter, H., Niedermeier, R., Sorge, M.: Adapting the Bron-Kerbosch algorithm for enumerating maximal cliques in temporal graphs. Soc. Netw. Anal. Min. **7**, 1–16 (2017)
14. Hou, B., et al.: Efficient maximal clique enumeration over graph data. Data Sci. Eng. **1**, 219–230 (2016)
15. Lu, Y., Miao, Z., Sahraeian, P., Balasundaram, B.: On atomic cliques in temporal graphs. Optim. Lett. **17**(4), 813–828 (2023)
16. Malliaros, F.D., Giatsidis, C., Papadopoulos, A.N., Vazirgiannis, M.: The core decomposition of networks: theory, algorithms and applications. VLDB J. **29**(1), 61–92 (2020)
17. Marathe, M., Vullikanti, A.K.S.: Computational epidemiology. Commun. ACM **56**(7), 88–96 (2013)
18. Mastrandrea, R., Fournet, J., Barrat, A.: Contact patterns in a high school: a comparison between data collected using wearable sensors, contact diaries and friendship surveys. PLoS ONE **10**(9), e0136497 (2015)
19. Milano, M., Cannataro, M.: Network models in bioinformatics: modeling and analysis for complex diseases (2023)

20. Molter, H., Niedermeier, R., Renken, M.: Enumerating isolated cliques in temporal networks. In: Cherifi, H., Gaito, S., Mendes, J.F., Moro, E., Rocha, L.M. (eds.) COMPLEX NETWORKS 2019. SCI, vol. 882, pp. 519–531. Springer, Cham (2020). https://doi.org/10.1007/978-3-030-36683-4_42
21. Qin, H., Li, R.H., Wang, G., Qin, L., Cheng, Y., Yuan, Y.: Mining periodic cliques in temporal networks. In: 2019 IEEE 35th International Conference on Data Engineering (ICDE), pp. 1130–1141. IEEE (2019)
22. Rossi, R.A., Gleich, D.F., Gebremedhin, A.H., Patwary, M.M.A.: Fast maximum clique algorithms for large graphs. In: Proceedings of the 23rd International Conference on World Wide Web, pp. 365–366 (2014)
23. Sanei-Mehri, S.V., Das, A., Hashemi, H., Tirthapura, S.: Mining largest maximal quasi-cliques. ACM Trans. Knowl. Discov. Data (TKDD) **15**(5), 1–21 (2021)
24. Stehlé, J., et al.: High-resolution measurements of face-to-face contact patterns in a primary school. PLOS ONE **6**(8), e23176 (2011). https://doi.org/10.1371/journal.pone.0023176
25. Viard, J., Latapy, M., Magnien, C.: Revealing contact patterns among high-school students using maximal cliques in link streams. In: Proceedings of the 2015 IEEE/ACM International Conference on Advances in Social Networks Analysis and Mining 2015, pp. 1517–1522 (2015)
26. Wang, J., Zeng, Z., Zhou, L.: Clan: an algorithm for mining closed cliques from large dense graph databases. In: 22nd International Conference on Data Engineering (ICDE 2006), p. 73. IEEE (2006)
27. Wood, D.R.: An algorithm for finding a maximum clique in a graph. Oper. Res. Lett. **21**(5), 211–217 (1997)
28. Xiang, J., Guo, C., Aboulnaga, A.: Scalable maximum clique computation using mapreduce. In: 2013 IEEE 29th International Conference on Data Engineering (ICDE), pp. 74–85. IEEE (2013)
29. Yang, S., Keller, F.B., Zheng, L.: Social Network Analysis: Methods and Examples. Sage Publications (2016)
30. Zhu, K., Fletcher, G., Yakovets, N., Papapetrou, O., Wu, Y.: Scalable temporal clique enumeration. In: Proceedings of the 16th International Symposium on Spatial and Temporal Databases, pp. 120–129 (2019)

Influential Billboard Slot Selection Under Zonal Influence Constraint

Dildar Ali, Suman Banerjee(✉), and Yamuna Prasad

Department of Computer Science and Engineering, Indian Institute of Technology Jammu, Jammu 181221, Jammu and Kashmir, India
{2021rcs2009,suman.banerjee,yamuna.prasad}@iitjammu.ac.in

Abstract. Given *billboard* and *trajectory database*, finding a limited number of billboard slots for maximizing the influence is an important problem in the context of *billboard advertisement*. Most of the existing literature focused on the influential slot selection problem without considering any specific zonal influence constraint. To bridge this gap in this paper, we introduce and study the Influential Billboard Slot Selection Problem Under Zonal Influence Constraint. We propose a simple greedy approach to solve this problem. Though this method is easy to understand and simple to implement due to the excessive number of marginal gain computations, this method is not scalable. We design a branch and bound framework with two bound estimation techniques that divide the problem into different zones and integrate the zone-specific solutions to obtain a solution for the whole. We implement both the solution methodologies with real-world billboard and trajectory datasets and several experiments have been reported. We compare the performance of the proposed solution approaches with several baseline methods. The results show that the proposed approaches lead to more effective solutions with reasonable computational overhead than the baseline methods.

Keywords: Billboard Database · Trajectory Database · Zonal Influence · Budget · Influence Maximization

1 Introduction

Billboard Advertising has become effective out-of-home marketing strategy, offering excellent returns on investment[1]. Several commercial houses have emerged as effective players in this domain. This area has grown tremendously in the last decade, and by the end of 2024, it has become 410.82 billion dollar industry[2]. To make an effective billboard advertisement, it is always important to choose

[1] https://topmediadvertising.co.uk/billboard-advertising-statistics/.
[2] https://www.thebusinessresearchcompany.com/report/billboard-and-outdoor-advertising-global-market-report.

J. Tekli et al. (Eds.): ADBIS 2024, LNCS 14918, pp. 93–106, 2024.
https://doi.org/10.1007/978-3-031-70626-4_7

most influential billboard slots that have reach to a large number of audiences. However, each billboard slot is associated with some cost, which means if a commercial house hires a particular slot, a pre-specified amount of money needs to be paid to the influence provider. As per the marketing literature reports, a commercial house spends around 7–10% of its annual revenue for advertising purposes[3]. So, it is an important question: given the information about billboard slots and corresponding costs, how can we select a limited number of them? This problem is called the BILLBOARD SLOT SELECTION Problem.

Background. To the best of our knowledge, Zhang et al. [9] were the first to study the billboard advertisement problem. Further, Zhang et al. [11] find a set of billboards with maximum influence under a budget when impression counts are considered. Zhang et al. [10] further extend their previous work [9] to find a set of most influential billboards under a budget constraint. Wang et al. [7] find k best advertising units which maximize the influence spread in targeted advertising. Similarly, Wang et al. [6] study the problem of targeted influence maximization under a limited budget constraint. Deviating from the influence maximization, some literature exists [3–5, 12] on regret minimization in the context of billboard advertisements. None of the studies have considered the budget associated with a billboard slot, unlike the study by Ali et al. [1,2]. Also, none of the studies have jointly considered the budget and zonal influence constraints. Now, we describe the motivation behind this constraint.

Motivation. In all the previous studies [1,9–11], It is assumed that each zone in a city is equally meaningful to an advertiser during the selection of slots. However, this is not always true in real-life scenarios [3]. Let's consider the zone-specific population information of a city is known to the advertiser. Now, it is meaningless to advertise a highly expensive product in a demographic zone where economically backward classes people live. Because they may not buy the product, investing money to advertise such a product by the advertiser is totally waste. Hence, we divide the city into several zones based on the demographic information. Now, the advertisers have zonal influence demand along with the budget, and the influence provider must fulfill this. So, studying the BILLBOARD SLOT SELECTION PROBLEM under zonal influence constraint is important.

Our Contributions. This paper introduces the problem of selecting influential billboard slots under zonal influence constraint. We call this problem INFLUENTIAL BILLBOARD SLOT SELECTION UNDER ZONAL INFLUENCE CONSTRAINT (IBSSP). To the best of our knowledge, this is the first study in this direction. The key contributions of this paper are as follows:

- We introduce and formulate this problem as a discrete optimization problem and show this problem as NP-hard to approximate within a constant factor.
- We propose a simple greedy and branch and bound approach with two bound estimation techniques to solve this problem.
- We substantiate our claim regarding the efficiency and effectiveness of the proposed solutions by experimental evaluations with real-world datasets.

[3] https://www.lamar.com/howtoadvertise/Research/.

Organization of the Paper. The rest of the paper is organized as follows. Section 2 describes the background of our problem and defines it formally. The proposed solution approaches have been described in Sect. 3. Section 4 contains the experimental evaluation of the proposed solution approaches. Finally, Sect. 5 concludes our study and mentions future research directions.

2 Background and Problem Definition

Consider a set of m digital billboards $\mathbb{B} = \{b_1, b_2, \ldots, b_m\}$ are placed across a geographical region of interest. Each one is operated for the duration $[T_1, T_2]$ and we denote $T = T_2 - T_1$. Each billboard is allocated slot-wise with the duration of the slot equal to Δ. Now, we define the notion of a billboard slot.

Definition 1 (Billboard Slot). *Given a set of billboards and a fixed duration of slot Δ, a billboard slot is defined by a tuple $(bs_i, [t, t+\Delta])$ where $bs_i \in \mathbb{B}$ and $t \in \{T_1, T_1+\Delta+1, T_1+2\Delta+1, \ldots, T_2-\Delta+1\}$.*

It can be observed that the number of billboard slots will be equal to $m \cdot \frac{T}{\Delta}$. Now, each billboard slot is associated with some cost, and this is formalized by the cost function $\mathcal{C}$ that maps each billboard slot to its associated cost $\mathcal{C}(bs_i)\ \forall\ bs_i \in \mathcal{BS}$, i.e., $\mathcal{C} : \mathcal{BS} \longrightarrow \mathbb{Z}^+$. Now, a subset of the slots can be chosen with a fixed budget of $\mathcal{B}$. Next, we define the influence of a subset of billboard slots in Definition 2.

Definition 2 (Influence of Billboard Slots). *Given a trajectory database $\mathbb{T}$, billboard database $\mathbb{B}$, and a subset of billboard slots $\mathcal{S} \subseteq \mathcal{BS}$, the influence of the billboard slots is given by Eq. 1.*

$$\mathcal{I}(\mathcal{S}) = \sum_{u_j \in \mathbb{T}} [1 - \prod_{bs_i \in \mathcal{S}} (1 - Pr(bs_i, u_j))] \tag{1}$$

Here, $\mathcal{I}()$ is the influence function that maps all possible subsets of the billboard slots to its expected influence, i.e., $\mathcal{I} : 2^{\mathcal{BS}} \longrightarrow \mathbb{R}_0^+$ with $\mathcal{I}(\emptyset) = 0$. $Pr(bs_i, u_j)$ denotes the probability of user u_j can be influenced by billboard slot bs_i. Now, consider the geographical region under consideration is denoted by $\mathcal{Z}$, and this is divided into ℓ many zones $\mathcal{Z} = \{z_1, z_2, \ldots, z_\ell\}$. Consider an advertiser a_i wants to choose a subset of billboard slots within the allocated budget $\mathcal{B}$ such that zonal influence demand is satisfied. Let, for any zone $z_i \in \mathcal{Z}$ the influence requirement is σ_i. The influence of the billboard slot subset $\mathcal{S}$ subject to the zone z_i is denoted by $\mathcal{I}_{z_i}(\mathcal{S})$ and $\mathcal{I}(\mathcal{S}) = \sum_{z_i \in \mathcal{Z}} \mathcal{I}_{z_i}(\mathcal{S})$. So the problem becomes: "Given a billboard database $\mathbb{B}$, trajectory database $\mathbb{T}$ and a cost function $\mathcal{C}$, the problem of Influential Billboard Slot Selection with Zonal Influence Constraint asks to choose a subset of the billboard slots $\mathcal{S}$ such that the total selection cost is less than the budget, i.e., $\sum_{b_i \in \mathcal{S}} \mathcal{C}(b_i) \leq \mathcal{B}$ and the zonal influence constraint for all the zones are satisfied, i.e., $\mathcal{I}_{z_i}(\mathcal{S}) \geq \sigma_i$ for all $i \in [\ell]$ such that the influence is maximized". Now, we define the zonal influence constraint in Definition 3.

Definition 3 (Zonal Influence Constraint). *Let $\mathbb{BS}_{a_i}$ be the set of allocated billboard slots to the advertiser a_i and $\mathbb{BS}_{a_i}^{z_j} \subseteq \mathbb{BS}_{a_i}$, contains the billboard slots associated with zone z_j. Now, we can say this allocation satisfies the zonal influence constraint for zone z_j if and only if $\mathcal{I}(\mathbb{BS}_{a_i}^{z_j}) \geq \sigma_i^j$.*

Next, this problem can be formally stated as follows:

SLOT SELECTION IN BILLBOARD ADVERTISEMENT Problem
Input: Billboard Slots set $\mathcal{BS}$, Trajectory Database $\mathbb{T}$, Influence Function $\mathcal{I}()$, Budget $\mathcal{B}$, Advertiser Database $\mathcal{A}$. **Problem:** Find out a subset of billboard slots $\mathcal{P} \subseteq \mathcal{BS}$, such that zonal influence demand is satisfied and total influence is maximized.

The influence maximization problem without considering zonal influence constraints in billboard advertisement has been studied by Zhang et al. [10], and they had inapproximability results stated in Theorem 1.

Theorem 1. THE INFLUENTIAL BILLBOARD SLOT SELECTION PROBLEM UNDER ZONAL INFLUENCE CONSTRAINT *is NP-hard and hard to approximate with a constant factor algorithm.*

This means the same inapproximability result also holds for our problem. Next, in Sect. 3, we discuss the proposed solution methodologies.

3 Proposed Solution Approach

3.1 Simple Greedy Allocation Scheme

This kind of greedy approach has already been studied by Zhang et al. [10] for influential billboard selection problems. Hence, we have extended the greedy approach for our problem context, i.e., under zonal influence constraint. A simple straight forward greedy approach for selecting billboard slots is defined in Algorithm 1 (Line No. 1 to 22), which selects the billboard slots that maximize the marginal influence gain to a candidate solution set $\mathcal{P}$ to fulfill the zone-specific influence demand of an advertiser until the budget ($\mathcal{B}$) is exhausted. However, a simple greedy approach does not guarantee the approximation ratio due to budget constraints. Hence, Line No. 23 to 49 is added with simple greedy to overcome the drawback and guarantees an approximation ratio. Basically, the naive greedy approach is used where the billboard slots are chosen based on the individual's highest influence value to solve the IBSSP problem.

Complexity Analysis. In-Line No. 1 initializing an empty sets $\mathcal{K}$,$\mathcal{P}$,$\mathcal{Q}$ will take $\mathcal{O}(1)$ time. In-Line No. 2 `for loop` and Line No. 4 `while loop` will execute for $\mathcal{O}(\ell)$ and $\mathcal{O}(m)$ time, respectively. In-Line No. 5, computing influence will take $\mathcal{O}(2 \cdot \ell \cdot m \cdot t)$ where t is the number of tuples in the trajectory database. Line No. 6 to 10 will execute for $\mathcal{O}(\ell \cdot m)$ time. Hence, Line No. 1 to 14 will take

Algorithm 1: Simple Greedy Approach for Influential Billboard Slot Selection Problem

Data: Trajectory Database $\mathbb{T}$, Billboard Database $\mathbb{B}$, Advertiser Database $\mathcal{A}$, and the Influence Function $\mathcal{I}()$, Zone Information, Budget $\mathcal{B}$, Zonal influence demand σ.
Result: A Billboard slot set $\mathcal{P} \subseteq \mathbb{BS}$ such that $\mathcal{C}(\mathcal{P}) \leq \mathcal{B}$

1 Initialize $\mathcal{K} \leftarrow \emptyset$, $\mathcal{P} \leftarrow \emptyset$, $\mathcal{Q} \leftarrow \emptyset$, $\mathcal{B}_1 = \mathcal{B}_2 = \mathcal{B}$;
2 **for** *each* $z_j \in a_z$ **do**
3 $\quad \mathcal{S} \leftarrow \emptyset$;
4 $\quad$ **while** $\sigma_{z_j} > I(\mathcal{S})$ *and* $\mathcal{BS}_{z_j} \neq \emptyset$ **do**
5 $\quad\quad s^* \leftarrow \underset{s \in \mathcal{BS}_{z_j}}{\text{argmax}} \frac{\mathcal{I}(\mathcal{S} \cup \{s\}) - \mathcal{I}(\mathcal{S})}{\mathcal{C}(\{s\})}$;
6 $\quad\quad$ **if** $\mathcal{C}(\mathcal{S}) + \mathcal{C}(\{s^*\}) \leq \mathcal{B}_1$ **then**
7 $\quad\quad\quad \mathcal{S} \leftarrow \mathcal{S} \cup \{s^*\}$;
8 $\quad\quad\quad \mathcal{BS}_{z_j} \leftarrow \mathcal{BS}_{z_j} \setminus \{s^*\}$;
9 $\quad\quad\quad \mathcal{B}_1 \leftarrow \mathcal{B}_1 - \mathcal{C}(s^*)$;
10 $\quad\quad$ **end**
11 $\quad$ **end**
12 $\quad \mathcal{P} \leftarrow \mathcal{P} \cup \{\mathcal{S}\}$;
13 **end**
14 $\mathcal{BS} \leftarrow \{\mathcal{BS}_{z_1} \cup \mathcal{BS}_{z_2} \cup \ldots \cup \mathcal{BS}_{z_l}\}$;
15 **while** $\mathcal{B}_1 > 0$ *and* $\mathcal{BS} \neq \emptyset$ **do**
16 $\quad s^* \leftarrow \underset{s \in \mathcal{BS}}{\text{argmax}} \frac{\mathcal{I}(\mathcal{S} \cup \{s\}) - \mathcal{I}(\mathcal{S})}{\mathcal{C}(\{s\})}$;
17 $\quad$ **if** $\mathcal{C}(\mathcal{P}) + \mathcal{C}(\{s^*\}) \leq \mathcal{B}_1$ **then**
18 $\quad\quad \mathcal{P} \leftarrow \mathcal{P} \cup \{s^*\}$;
19 $\quad\quad \mathcal{BS} \leftarrow \mathcal{BS} \setminus \{s^*\}$;
20 $\quad\quad \mathcal{B}_1 \leftarrow \mathcal{B}_1 - \mathcal{C}(s^*)$;
21 $\quad$ **end**
22 **end**
23 **for** *each* $z_j \in a_z$ **do**
24 $\quad \mathcal{K} \leftarrow \emptyset$;
25 $\quad$ **while** $\sigma_{z_j} > I(\mathcal{K})$ *and* $\mathcal{BS}_{z_j} \neq \emptyset$ **do**
26 $\quad\quad k^* \leftarrow \underset{s \in \mathcal{BS}_{z_j}}{\text{argmax}} \mathcal{I}(\mathcal{K} \cup \{s\})$;
27 $\quad\quad$ **if** $\mathcal{C}(\mathcal{K}) + \mathcal{C}(\{k^*\}) \leq \mathcal{B}_2$ **then**
28 $\quad\quad\quad \mathcal{K} \leftarrow \mathcal{K} \cup \{k^*\}$;
29 $\quad\quad\quad \mathcal{BS}_{z_j} \leftarrow \mathcal{BS}_{z_j} \setminus \{k^*\}$;
30 $\quad\quad\quad \mathcal{B}_2 \leftarrow \mathcal{B}_2 - \mathcal{C}(k^*)$;
31 $\quad\quad$ **end**
32 $\quad$ **end**
33 $\quad \mathcal{Q} \leftarrow \mathcal{Q} \cup \{\mathcal{K}\}$;
34 **end**
35 $\mathcal{BS} \leftarrow \{\mathcal{BS}_{z_1} \cup \mathcal{BS}_{z_2} \cup \ldots \cup \mathcal{BS}_{z_l}\}$;
36 **while** $\mathcal{B}_2 > 0$ *and* $\mathcal{BS} \neq \emptyset$ **do**
37 $\quad k^* \leftarrow \underset{s \in \mathcal{BS}}{\text{argmax}} \mathcal{I}(\mathcal{K} \cup \{s\})$;
38 $\quad$ **if** $\mathcal{C}(\mathcal{Q}) + \mathcal{C}(\{k^*\}) \leq \mathcal{B}_2$ **then**
39 $\quad\quad \mathcal{Q} \leftarrow \mathcal{Q} \cup \{k^*\}$;
40 $\quad\quad \mathcal{BS} \leftarrow \mathcal{BS} \setminus \{k^*\}$;
41 $\quad\quad \mathcal{B}_2 \leftarrow \mathcal{B}_2 - \mathcal{C}(k^*)$;
42 $\quad$ **end**
43 **end**
44 **if** $\mathcal{I}(\mathcal{Q}) > \mathcal{I}(\mathcal{P})$ **then**
45 $\quad$ return $\mathcal{Q}$;
46 **end**
47 **else**
48 $\quad$ return $\mathcal{P}$
49 **end**

$\mathcal{O}(m^2 \cdot \ell \cdot t^2)$. Similarly, Line No. 15 to 22 will take $\mathcal{O}(m^2 \cdot t)$. In Line No. 23 to 35 will take $\mathcal{O}(m^2 \cdot \ell \cdot t^2)$ and Line No. 36 to 49 will take $\mathcal{O}(m^2 \cdot t)$ time to execute. So, Algorithm 1 will take total $\mathcal{O}(m^2 \cdot \ell \cdot t^2 + m^2 \cdot t)$ i.e., $\mathcal{O}(m^2 \cdot \ell \cdot t^2)$ time to execute. The extra space taken by Algorithm 1 will be of $\mathcal{O}(\ell + m)$.

3.2 Branch and Bound Approach

Although Simple Greedy achieves a constant approximation guarantee, executing requires huge amounts of computational time. To overcome the drawback, we introduce the Branch and Bound approach in Algorithm 2. We first initialize U_G, L_G and the max heap $(\mathcal{P}, \bar{\mathcal{P}}, \mathcal{Y}, \mathcal{U})$, where $\mathcal{P}$ is the solution set, $\bar{\mathcal{P}}$ is the set of unexplored billboard slots, $\mathcal{Y}$ is the zone-specific influence demand, and $\mathcal{U}$ is the upper bound of the influence. The max heap $\mathcal{H}$ is ordered based on the upper bound of influence $\mathcal{U}$. Hence, the heap will pop the top entry, which has a maximum $\mathcal{U}$ value till $L_G < U_G$. Further, as long as the budget constraint is satisfied, it will generate two new branches $\mathcal{P}^a$ and $\mathcal{P}^b$. Initially, $\mathcal{P}^a$ denotes the feasible slots set by adding b while $\mathcal{P}^b$ is the feasible slots set excluding b (Lines 9 - 11). Now, based on the value of $\mathcal{P}^a$ (or $\mathcal{P}^b$) and $\bar{\mathcal{P}}$ value the $BoundEstimation(.)$ or $FastBoundEstimation$ function will return $(\mathcal{P}^c, L^a, \mathcal{Y}, \mathcal{U}^a)$ or $(\mathcal{P}^c, L^b, \mathcal{Y}, \mathcal{U}^b)$.

In-Line No. 13 if the current satisfied influence demand (L^a) is greater than previous satisfied demand (L^G) then update L^G with L^a and $\hat{\mathcal{P}}$ with $\mathcal{P}^c$ as it may be the possible solution set. Next, in Line No. 16, if updated influence demand ($\mathcal{U}^a$) still greater than L_G, then insert that node into the max heap $\mathcal{H}$ for further explorations. We do the same in Line No. 19 to 25 for $\mathcal{P}^b$. We repeat the `while loop` for all the branches until $L_G \geq U_G$.

Algorithm 2: Branch and Bound Approach for Influential Billboard Slot Selection Problem

---

Data: Trajectory Database $\mathbb{T}$, Billboard Database $\mathbb{B}$, and the Influence Function $I()$, Zonal Demand $\mathcal{Y}$, Zone Info. a_z, Budget $\mathcal{B}$.
Result: A Billboard slot set $\hat{\mathcal{P}} \subseteq \mathbb{BS}$ such that $\mathcal{C}(\hat{\mathcal{P}}) \leq \mathcal{B}$
1 Initialize $\hat{\mathcal{P}} \leftarrow \emptyset$, $\mathcal{P} \leftarrow \emptyset$, $\bar{\mathcal{P}} \leftarrow \mathbb{BS}, \mathcal{Y} = (\sigma_{z_1}, \sigma_{z_2}, \ldots \sigma_{z_\ell})$;
2 Initialize max heap $\mathcal{H} \leftarrow (\mathcal{P}, \bar{\mathcal{P}}, \mathcal{Y}, \mathcal{U})$;
3 $L_G \leftarrow 0$, $U_G \leftarrow \infty$;
4 **while** $L_G < \theta . U_G$ **do**
5     $(\mathcal{P}, \bar{\mathcal{P}}, \mathcal{Y}, \mathcal{U}) \leftarrow$ top of $\mathcal{H}$;
6     $U_G \leftarrow \mathcal{U}$;
7   **for** *each* $b \in \bar{\mathcal{P}}$ **do**
8   **if** $\mathcal{C}(\mathcal{P}) + \mathcal{C}(b) \leq \mathcal{B}$ **then**
9             $\bar{\mathcal{P}} \leftarrow \bar{\mathcal{P}} \setminus \{b\}$;
10             $\mathcal{P}^a \leftarrow \mathcal{P} \cup \{b\}$;
11             $\mathcal{P}^b \leftarrow \mathcal{P}$;
12             $(\mathcal{P}^c, L^a, \mathcal{Y}, \mathcal{U}^a) \leftarrow$ BoundEstimation$(\mathcal{P}^a, \bar{\mathcal{P}}, a_z, \mathcal{Y})$;
13   **if** $L^a > L_G$ **then**
14                 $L_G \leftarrow L^a$, $\hat{\mathcal{P}} \leftarrow \mathcal{P}^c$;
15   **end**
16   **if** $\mathcal{U}^a > L_G$ **then**
17                 $\mathcal{H} \leftarrow \mathcal{H} \cup (\mathcal{P}^a, \bar{\mathcal{P}}, \mathcal{Y}, \mathcal{U}^a)$;
18   **end**
19             $(\mathcal{P}^c, L^b, \mathcal{Y}, \mathcal{U}^b) \leftarrow$ BoundEstimation$(\mathcal{P}^b, \bar{\mathcal{P}}, a_z, \mathcal{Y})$;
20   **if** $L^b > L_G$ **then**
21                 $L_G \leftarrow L^b$, $\hat{\mathcal{P}} \leftarrow \mathcal{P}^c$;
22   **end**
23   **if** $\mathcal{U}^b > L_G$ **then**
24                 $\mathcal{H} \leftarrow \mathcal{H} \cup (\mathcal{P}^b, \bar{\mathcal{P}}, \mathcal{Y}, \mathcal{U}^b)$;
25   **end**
26   **end**
27   **end**
28 **end**
29 return $\hat{\mathcal{P}}$;

Complexity Analysis. Now, we analyze the time and space requirements for Algorithm 2. Initialization at Line No. 1, 3 will take $\mathcal{O}(1)$ time and initialize max heap in Line No. 2 will take $\mathcal{O}(m \cdot \log \cdot m)$ time. In-Line No. 4 `while loop` will execute for $\mathcal{O}(n)$ till $L^G < U^G$. To pick an element in Line No. 7 will take $\mathcal{O}(1)$, and Line No. 8 to 11 will take $\mathcal{O}(m)$ time to execute. In-Line No. 12 for each slot, it will take $\mathcal{O}(\ell \cdot m^2 \cdot t)$ time to execute. Next, in Line No. 13 to 15 will take $\mathcal{O}(1)$ time and Line No. 16 to 18 will take $\mathcal{O}(m \cdot \log \cdot m)$ time. So, Line No. 12 to 18 will take $\mathcal{O}(\ell \cdot m^2 \cdot t + m \cdot \log \cdot m)$ time to execute. Similarly, Line No. 19 to 25 will take $\mathcal{O}(\ell \cdot m^2 \cdot t + m \cdot \log \cdot m)$ time in the worst case. Hence, In Algorithm 2 will take total $\mathcal{O}(\ell \cdot m^2 \cdot t + m \cdot \log \cdot m)$ time to execute the `while loop` each time. The additional space requirements for Algorithm 2 will be of $\mathcal{O}(m + \ell)$.

3.3 Fast Bound Estimation

The Fast Bound Estimation approach is introduced in Algorithm 3 to generate and explore the branches in each step. A key challenge is to find a tight upper bound due to the overlap influence of the billboard slots. So, to achieve this goal, we first allocate billboard slots $\mathcal{P}^c = \{b_1, b_2, \dots b_k\}$, where $k \in \mathcal{BS}$ based on the highest influence value till the budget constraint is satisfied. Let's assume b^{extra} is the next billboard slot with the highest influence value. Now, if we include b^{extra} into $\mathcal{P}^c$ total budget is exceeded, but if not included, we will lose the cost $\mathcal{B} - \mathcal{C}(\mathcal{P}^c)$. Therefore we add b^{extra} to get the upper bound, and it is calculated as $\mathcal{U}^a \leftarrow \mathcal{I}(\mathcal{P}^c) + [\mathcal{B} - \mathcal{C}(\mathcal{P}^c)].\frac{\mathcal{I}(\mathcal{P}^c \cup\ b^{extra}) - \mathcal{I}(\mathcal{P}^c)}{\mathcal{C}(b^{extra})}$. This Algorithm starts by initializing p^* and zone-specific influence demand $\mathcal{Y}$. In-Line No. 2 `for loop` will execute for all the zones an advertiser has, and in Line, No. 3 the `while loop` will run until the budget is exhausted and zonal influence demands are satisfied. If the budget and remaining influence demand are greater than zero, then Line No. 14 to 21 will be executed to fulfill the demand. In-Line No. 23 to 24, the lower and upper bound is calculated, and in Line No. 25 $\mathcal{P}^c$, $\mathcal{Y}$, $\mathcal{U}^a$ and L^a is returned.

Algorithm 3: FastBoundEstimation

Data: $\mathcal{D}$, $\mathbb{BS}$, $I()$, z_j, $\mathcal{B}$, $\mathcal{P}^a$, $\bar{\mathcal{P}}$, $\mathcal{Y}$.
Result: $(\mathcal{P}^c, L^a, \mathcal{Y}, \mathcal{U}^a)$

1 $p^* \leftarrow \emptyset$, $\mathcal{P}^c \leftarrow \emptyset$;
2 **for** *each* $z_j \in a_z$ **do**
3   **while** $\mathcal{C}(\mathcal{P}^a) + \mathcal{C}(p^*) \leq \mathcal{B}$ *and* $\mathcal{I}(\mathcal{P}^a \cup \{p^*\}) \leq \sigma_{z_j}$ *and* $|\bar{\mathcal{P}}| \neq \emptyset$ **do**
4         $b \leftarrow \underset{s \in \bar{\mathcal{P}}}{argmax}\ \mathcal{I}((\mathcal{P}^a \cup \{p^*\}) \cup \{s\})$;
5   **if** $\mathcal{C}(\mathcal{P}^a \cup p^*) + \mathcal{C}(b) \leq \mathcal{B}$ **then**
6             $p^* \leftarrow p^* \cup \{b\}$;
7             $\mathcal{B} \leftarrow \mathcal{B} - \mathcal{C}(b)$;
8   **end**
9         $\bar{\mathcal{P}} \leftarrow \bar{\mathcal{P}} \setminus \{b\}$;
10   **end**
11     $\mathcal{P}^c \leftarrow \mathcal{P}^c \cup (\mathcal{P}^a \cup p^*)$;
12 **end**
13 $p^* \leftarrow \emptyset$;
14 **while** $\mathcal{C}(\mathcal{P}^c) + \mathcal{C}(p^*) \leq \mathcal{B}$ *and* $|\bar{\mathcal{P}}| \neq \emptyset$ **do**
15     $b \leftarrow \underset{s \in \bar{\mathcal{P}}}{argmax}\ \mathcal{I}((\mathcal{P}^a \cup \{p^*\}) \cup \{s\})$;
16   **if** $\mathcal{C}(\mathcal{P}^c \cup p^*) + \mathcal{C}(b) \leq \mathcal{B}$ **then**
17         $p^* \leftarrow p^* \cup \{b\}$;
18         $\mathcal{B} \leftarrow \mathcal{B} - \mathcal{C}(b)$;
19   **end**
20     $\bar{\mathcal{P}} \leftarrow \bar{\mathcal{P}} \setminus \{b\}$;
21 **end**
22 $\mathcal{P}^c \leftarrow \mathcal{P}^c \cup p^*$, $\bar{\mathcal{P}} \leftarrow \bar{\mathcal{P}} \setminus \mathcal{P}^c$;
23 $b^{extra} \leftarrow \underset{s \in \bar{\mathcal{P}}}{argmax}\ \mathcal{I}(\mathcal{P}^c \cup \{s\})$;
24 $L^a \leftarrow \mathcal{I}(\mathcal{P}^c)$, $\mathcal{U}^a \leftarrow \mathcal{I}(\mathcal{P}^c) + [\mathcal{B} - \mathcal{C}(\mathcal{P}^c)].\frac{\mathcal{I}(\mathcal{P}^c \cup\ b^{extra}) - \mathcal{I}(\mathcal{P}^c)}{\mathcal{C}(b^{extra})}$;
25 return $\mathcal{P}^c, L^a, \mathcal{Y}, \mathcal{U}^a$

Complexity Analysis. Initialization at Line No. 1 will take $\mathcal{O}(1)$ time to execute. The `for loop` at Line No. 2 will execute for $\mathcal{O}(\ell)$ time and `while loop` at line No. 3 will execute for $\mathcal{O}(m)$ time and in Line No. 4, computing marginal gain will take $\mathcal{O}(\ell \cdot m^2 \cdot t)$ time in which t is the number of tuples in the trajectory database. Line No. 5 to 11 will execute for $\mathcal{O}(m \cdot \ell)$ time. Hence, Line No. 1 to 12 will take total $\mathcal{O}(\ell \cdot m^2 \cdot t + m \cdot \ell)$ time to execute. Similarly, Line No. 13 to 22 will take $\mathcal{O}(\ell \cdot m^2 \cdot t + m \cdot \ell)$ time in the worst case. In-Line No. 23, calculating b^{extra} will take $\mathcal{O}(m \cdot t)$ and Line No. 24 will take $\mathcal{O}(4 \cdot m \cdot t)$ time. Hence, Algorithm 3 will take $\mathcal{O}(\ell \cdot m^2 \cdot t + m \cdot \ell + m \cdot t)$ i.e., $\mathcal{O}(\ell \cdot m^2 \cdot t)$ time in the worst case. The additional space requirements for Algorithm 3 will be $\mathcal{O}(m + \ell)$.

3.4 Bound Estimation

The branch and bound approach invokes Algorithm 4 for bound estimation in each iteration. From Algorithm 1, we observed that selecting billboard slots using greedy takes huge computational time to calculate the marginal gain and find the slots having the maximum gain. Although the branch and bound with Algorithm 3 gives a solution based on the highest influence value in each iteration, it is not optimal. However, it reduces computational costs hugely. Motivated by this, we propose a bound estimation approach in Algorithm 4 without traversing all the slots to estimate the bound with an approximation of $\frac{\theta}{2}(1 - 1/e - \epsilon)$, where ϵ is a user-defined parameter which balances the trade-off between accuracy and efficiency. Also, we set a threshold τ in Line No. 2 by calculating maximal of $(\mathcal{I}(\mathcal{P}^a \cup \{b\}) - \mathcal{I}(\mathcal{P}^a))/\mathcal{C}(b)$ and we progressively decrease the value of τ by a factor of $(1 + e)$ and insert an element into p^*. Here, one thing needs to be observed: Algorithm 4 does not explore all the billboard slots to find the best solution, and when the value of τ is small enough, the algorithm terminates.

Algorithm 4: BoundEstimation

Data: $\mathcal{D}$, $\mathbb{BS}$, $I()$, z_j, $\mathcal{B}$, $\mathcal{P}^a$, $\bar{\mathcal{P}}$, $\mathcal{Y}$.
Result: $(\mathcal{P}^c, L^a, \mathcal{Y}, \mathcal{U}^a)$

1 $p^* \leftarrow \emptyset$, $\mathcal{P}^c \leftarrow \emptyset$, $\epsilon \leftarrow 0.1$;
2 $\tau \leftarrow \max_{b \in \bar{\mathcal{P}}} \frac{\mathcal{I}(\mathcal{P}^a \cup \{b\}) - \mathcal{I}(\mathcal{P}^a)}{\mathcal{C}(b)}$, $\mathcal{B}^r \leftarrow \mathcal{B} - \mathcal{C}(\mathcal{P}^a)$;
3 **for** *each* $z_j \in a_z$ **do**
4   **while** $\mathcal{C}(\mathcal{P}^a) + \mathcal{C}(p^*) \leq \mathcal{B}$ *and* $\mathcal{I}(\mathcal{P}^a \cup \{p^*\}) \leq \sigma_{z_j}$ *and* $|\bar{\mathcal{P}}| \neq \emptyset$ **do**
5   **for** $b \in \bar{\mathcal{P}}$ **do**
6   **if** $\mathcal{C}(\mathcal{P}^a \cup p^*) + \mathcal{C}(b) \leq \mathcal{B}$ **then**
7         $m = \mathcal{I}(\mathcal{P}^a \cup \{b\} | \mathcal{P}^a) - \mathcal{I}(p^* | \mathcal{P}^a)$;
8   **if** $\frac{m}{\mathcal{C}(b)} \geq \tau$ **then**
9           $\bar{\mathcal{P}} \leftarrow \bar{\mathcal{P}} \setminus \{b\}$;
10           $p^* \leftarrow p^* \cup \{b\}$;
11   **end**
12   **if** $\frac{m}{\mathcal{C}(b)} < \tau$ **then**
13   break;
14   **end**
15   **end**
16   **end**
17     $\tau \leftarrow \frac{\tau}{1+\epsilon}$;
18   **if** $\tau \leq \frac{\mathcal{I}(p^* | \mathcal{P}^a)}{\mathcal{B}^r} \cdot \frac{e^{-1}}{1 - e^{-1}}$ **then**
19   break;
20   **end**
21   **end**
22   $\mathcal{P}^c \leftarrow \mathcal{P}^c \cup (\mathcal{P}^a \cup p^*)$;
23 **end**
24 $p^* \leftarrow \emptyset$, $\bar{\mathcal{P}} \leftarrow \bar{\mathcal{P}} \setminus \mathcal{P}^c$, $\mathcal{B} \leftarrow \mathcal{B} - \mathcal{C}(\mathcal{P}^c)$;
25 **while** $\mathcal{C}(\mathcal{P}^c) + \mathcal{C}(p^*) \leq \mathcal{B}$ *and* $|\bar{\mathcal{P}}| \neq \emptyset$ **do**
26   **for** $b \in \bar{\mathcal{P}}$ **do**
27   **if** $\mathcal{C}(\mathcal{P}^c \cup p^*) + \mathcal{C}(b) \leq \mathcal{B}$ **then**
28       $m = \mathcal{I}(\mathcal{P}^c \cup \{b\} | \mathcal{P}^c) - \mathcal{I}(p^* | \mathcal{P}^c)$;
29   **if** $\frac{m}{\mathcal{C}(b)} \geq \tau$ **then**
30         $\bar{\mathcal{P}} \leftarrow \bar{\mathcal{P}} \setminus \{b\}$;
31         $p^* \leftarrow p^* \cup \{b\}$;
32   **end**
33   **if** $\frac{m}{\mathcal{C}(b)} < \tau$ **then**
34   break;
35   **end**
36   **end**
37   **end**
38   $\tau \leftarrow \frac{\tau}{1+\epsilon}$;
39   **if** $\tau \leq \frac{\mathcal{I}(p^* | \mathcal{P}^a)}{\mathcal{B}^r} \cdot \frac{e^{-1}}{1 - e^{-1}}$ **then**
40   break;
41   **end**
42 **end**
43 $\mathcal{P}^c \leftarrow \mathcal{P}^c \cup p^*$, $\bar{\mathcal{P}} \leftarrow \bar{\mathcal{P}} \setminus \mathcal{P}^c$;
44 $b^{extra} \leftarrow \underset{s \in \bar{\mathcal{P}}}{argmax} \frac{\mathcal{I}(\mathcal{P}^c \cup \{s\}) - \mathcal{I}(\mathcal{P}^c)}{\mathcal{C}(\{s\})}$;
45 $L^a \leftarrow \mathcal{I}(\mathcal{P}^c)$, $\mathcal{U}^a \leftarrow \mathcal{I}(\mathcal{P}^c) + [\mathcal{B} - \mathcal{C}(\mathcal{P}^c)] . \frac{\mathcal{I}(\mathcal{P}^c \cup b^{extra}) - \mathcal{I}(\mathcal{P}^c)}{\mathcal{C}(b^{extra})}$;
46 return $\mathcal{P}^c, L^a, \mathcal{Y}, \mathcal{U}^a$

Complexity Analysis. Initialization at Line No. 1 and calculating τ at Line No. 2 will take $\mathcal{O}(1)$ and $\mathcal{O}(2 \cdot m \cdot t)$ time, respectively. Line No. 3 `for loop` and Line No. 4 `while loop` will execute for $\mathcal{O}(\ell)$ and $\mathcal{O}(m)$ time, respectively. Line No. 5 `for loop` will execute for $\mathcal{O}(m^2 \cdot \ell)$ and Line No. 7 will take $\mathcal{O}(2 \cdot m^2 \cdot \ell \cdot t)$ time to execute. Line No. 8 to 16 will execute for $\mathcal{O}(m^2 \cdot \ell)$ and Line No. 17 will execute for $\mathcal{O}(m \cdot \ell)$. Finally, in Line No. 18 to 20 will take $\mathcal{O}(m \cdot \ell \cdot t)$ time. Hence, Line No. 1 to 23 will take $\mathcal{O}(2 \cdot m \cdot t + 2 \cdot m^2 \cdot \ell \cdot t + m^2 \cdot \ell + m \cdot \ell + m \cdot \ell \cdot t)$ i.e., $\mathcal{O}(2 \cdot m^2 \cdot \ell \cdot t)$ time to execute. Similarly, Line No. 24 to 43 will take $\mathcal{O}(2 \cdot m^2 \cdot \ell \cdot t)$ time to execute. In-Line No. 44 will take $\mathcal{O}(m \cdot t)$ and Line No. 45 will take $\mathcal{O}(4 \cdot m \cdot t)$ time to execute. Hence, Algorithm 4 will take $\mathcal{O}(2 \cdot m^2 \cdot \ell \cdot t + 2 \cdot m^2 \cdot \ell \cdot t + 2 \cdot m \cdot t + 4 \cdot m \cdot t)$ i.e., $\mathcal{O}(m^2 \cdot \ell \cdot t)$ time in the worst case to execute. The additional space requirements for Algorithm 4 will be $\mathcal{O}(m + \ell)$.

Theoretical Analysis Related to Performance. Due to space limitations, we have not conducted an elaborate analysis of our proposed methods. In our case, we have used the results from the study by Zhang et al. [11] and state their result in Theorem 2.

Theorem 2. *The branch and bound approach invoking Algorithm 4 and it results an optimal solution set ($\mathcal{OPT}$) from the billboard slots set $\bar{\mathcal{P}}$. It follows the following criteria:*

$$\mathcal{I}(\bar{\mathcal{P}}) \geq \frac{\theta}{2}(1 - 1/e - \epsilon) \cdot \mathcal{I}(\mathcal{OPT}). \tag{2}$$

An Illustrative Example. Consider an influence provider has four billboard slots $\mathcal{BS} = \{b_1, b_2, b_3, b_4\}$ and an advertiser a_1 approaches to the influence provider with required influence demand from two different zones z_1 and z_2 for budget $\mathcal{B}$. It can be denoted in the form of vector $[(\sigma_{z_1}, \sigma_{z_2}), \mathcal{B}]$. Now, among these four slots b_1 and b_2 belongs to zone z_1 with influence value 2 and 3 respectively, b_3 belongs to zone z_2 with influence value 7 and b_4 belongs to zone z_3 with influence value 5 as shown in Fig. 1. Next, the working of Branch and Bound framework in Algorithm 2 is as follows: At first, we initialize the max heap $\mathcal{H}$ with an entry $(\{\}, \{b_1, b_2, b_3, b_4\}, [(5, 7), 1000])$. The set $\mathcal{P}$ is empty, and the set $\bar{\mathcal{P}}$ contains all the billboard slots $\{b_1, b_2, b_3, b_4\}$ in the top part of the search tree as shown in the Fig. 2. In the search tree, every node has an entry like $(\mathcal{P}, \bar{\mathcal{P}}, \mathcal{U})$ and in each step the values of $\mathcal{P}$, $\bar{\mathcal{P}}$, and $\mathcal{U}$ are updated. Next, in step 1, we pop the initial heap entry and, based on the budget and zonal constraint, push a new branch into the heap, i.e., $(\{b_1\}, \{b_2, b_3, b_4\}, [(3, 7), 900])$ by invoking Algorithm 3 or 4. Since the $L^a > L_G$, we update the value of $\hat{\mathcal{P}}$ with $\mathcal{P}^c$. Similarly, at steps $2, 3, 4$, the newly generated branches are pushed into the heap. However, in step $4, 7, 10, 13, 15, 17$, adding billboard slot b_4 does not satisfy the zonal influence constraint, and these entries are not inserted into the heap. Next, in step 14, the zonal influence demand is satisfied, but some budget remains. So, this budget is utilized by taking slots from available zones, i.e., b_4 is added to the solution set, and the budget is exhausted. From Fig. 2, it is clear that the path (*step* 1 $\rightarrow$ *step* 5 $\rightarrow$ *step* 14 $\rightarrow$ *step* 18) gives an optimal solution that satisfies all the constraints and when $L_G \geq U_G$ Algorithm 2 terminates.

4 Experimental Evaluation

This section discusses the experimental evaluation to show the effectiveness of the proposed solution methodologies. Initially, we start by describing the datasets.

$\mathcal{BS}_i$	bs_1	bs_2	bs_3	bs_4
$I(bs_i)$	2	3	7	5
$C(bs_i)$	100	200	400	300
Zone Info.	z_1	z_1	z_2	z_3

Fig. 1. Billboard Info.

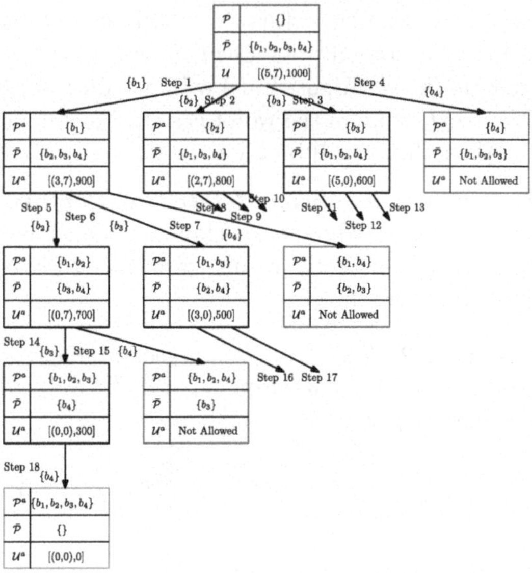

Fig. 2. A running example

Dataset Descriptions. We conduct experiments on real-world trajectory check-in datasets for New York City (NYC) and Los Angeles (LA). The NYC[4] dataset contains 227,428 check-ins between April 12, 2012, and February 16, 2013, and The LA[5] dataset contains 74,170 check-in from 15 different streets of Los Angles. We have crawled billboard datasets from the different locations of New York City and Los Angeles from LAMAR[6], one of the largest billboard providers worldwide. The NYC billboard dataset consists of 1031040 slots, and the LA dataset consists of 2135520 slots. However, in our experiment, we used 1500 and 2500 slots for NYC and LA, respectively. Additionally, we divide the billboard datasets based on latitude and longitude for New York City and Los Angeles into 5 and 3 different geographic zones, respectively.

4.1 Experimental Setup

The following experimental setups are considered in our experiments.

Key Parameters. All the key parameters are summarized in Table 1, and the default parameters are highlighted in bold. In our experiments, we vary only one parameter and keep the rest of the parameters in the default setting. In Algorithm 2, θ controls the termination condition while ϵ is used to make trade efficiency with accuracy. The distance threshold η denotes the maximum distance in which a billboard can influence the maximum number of trajectories.

4 https://www.nyc.gov/site/tlc/about/tlc-trip-record-data.page.
5 https://github.com/Ibtihal-Alablani.
6 http://www.lamar.com/InventoryBrowser.

Table 1. Key Parameters

Parameter	Values
$\mathcal{B}$	$5k, 10k, 15k, 20k, 25k$
ϵ	$10^{-4}, 10^{-3}, 10^{-2}, 10^{-1}, 1$
θ	$0.5, 0.6, \mathbf{0.7}, 0.8, 0.9$
η	$25m, 50m, \mathbf{100m}, 125m, 150m$

Environment Setup. All Python codes were executed on an HP Z4 workstation equipped with 64 GB of memory and an Xeon(R) 3.50 GHz processor.

Billboard Cost. An advertising company like LAMAR does not provide an exact cost of billboard slots. As reported in the existing studies [8,9,11], a billboard's cost is proportional to its corresponding influence value. Hence, we have calculated using the following setting: $\mathcal{C}(b) = \lfloor \delta * \mathcal{I}(b)/10 \rfloor$, where δ is a factor randomly chosen between 0.8 to 1.1 to simulate the cost fluctuation.

Performance Measurement. Each baseline and the proposed method are evaluated for their influence and runtime, and each experiment is repeated five times to demonstrate the effectiveness and efficiency of the proposed methods. Average results from these experiments are reported.

4.2 Algorithms Compared

In this paper, our goal is to evaluate the effectiveness and efficiency of the proposed methods. Hence, we compare our methods with the following baselines:

Top-k. Each iteration selects a billboard slot, which influences a maximum number of trajectories until the zone-specific influence demand and budget constraints are satisfied.

Random. This is the simple approach where it will choose billboard slots uniformly at random in each iteration until the zonal influence and budget constraints are satisfied.

4.3 Goals of the Experiments

In this study, we want to address the following Research Questions (RQ).

- **RQ1:** How do the influence value and runtime change if we increase the budget to select billboard slots?
- **RQ2:** For the Branch and Bound algorithm, varying values of θ and ϵ, how do the runtime and the influence quality change?
- **RQ3:** How do the influence value and run-time change if the number of demanding zones increases by the advertisers?
- **RQ4:** How do the influence and run-time change if we vary trajectory size?

4.4 Experimental Results

This section will discuss the experimental results of the solution methodologies and address research questions mentioned in Sect. 4.3.

Budget vs. Influence. Figure 3(a) and 3(c) show the effectiveness of the proposed as well as baseline methodologies in the NYC and LA datasets. From this, we have significant observations. First, when the budget increases from $5k$ to $25k$ in NYC and $50k$ to $150k$ in the LA dataset, the BBS outperforms the BFBS and the greedy approach by 26% to 28% and 20% to 27% respectively. Second, the BFBS has worse performance, as it gives preference to the slots with the highest influence value in each iteration, which are the most expensive in the real world. So, it can choose only a limited number of slots when the budget is less, and due to the influence overlap, the BBS performs well compared to the BFBS approach. Third, the impact of the BBS and the BFBS is greater in NYC than in the LA dataset, and the possible reason is the distribution of trajectory in the trajectory dataset, i.e., the influence overlap is less in the NYC Dataset.

Budget vs. Time. From Fig. 3(b) and 3(d), we have three observations. First, the running time of BFBS is very low compared to the BBS and Greedy approach because in BFBS, for every branch, the slots are selected based on the highest individual influence value, whereas in Greedy, slots are selected based on marginal gain computation. However, for every branch, the BBS method compute marginal gain for each slot in each iteration, which resulted in a significant increase in run time. Second, with the increased budget, the computational cost of each proposed and baseline method increases. Third, Top-k and Random is the fastest approach because they scan all billboard slots only once.

Varying ϵ,θ. The user-defined parameter θ controls the termination criteria in Algorithm 2. The larger θ value will increase the effectiveness with worse efficiency. Figure 3(c), 3(d) shows the effect of varying θ in the NYC dataset. Next, when the ϵ value increases from 0.0001 to 1, the effectiveness of the BFBS and BBS decreases by at least 5% to 10%.However, the efficiency increases by 20% to 30%. We found that the effectiveness and efficiency are stable when θ and ϵ values are 0.7 and 0.1, respectively, and chose this as the default setting.

Scalability Test. Figure 3(g) shows that in the NYC dataset, efficiency is very sensitive in BBS compared to the BFBS when we vary the billboard slot size from 300 to 1500. It is observed that the run-time increase rate in BBS compared to Greedy and BFBS is almost 57% to 86% and 92% to 95%, respectively.

Additional Discussion. Now, we discuss the impact of additional parameters, e.g., varying distance (η), no. of zones ($|\ell|$), size of the trajectory ($|\mathcal{T}|$), etc. First, the influence of all the proposed and baseline methods increases with the distance increase. This happens because each billboard slot can influence a large number of trajectories. Second, when zone-specific influence demand rises with the fixed budget, the computational cost of BBS and BFBS decreases. This happens because most of the budget is used to select slots for demanding zones, and very little is used to choose slots from the remaining slots in any of the

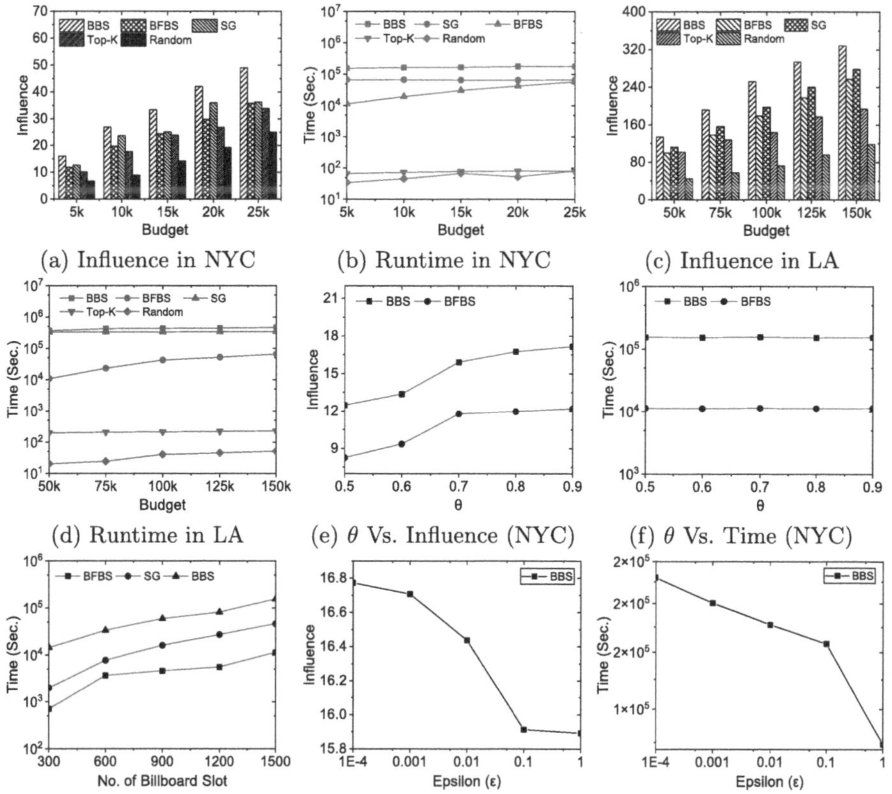

(a) Influence in NYC (b) Runtime in NYC (c) Influence in LA

(d) Runtime in LA (e) θ Vs. Influence (NYC) (f) θ Vs. Time (NYC)

(g) Scalability Test (NYC) (h) ϵ Vs. Influence (NYC) (i) ϵ Vs. Time (NYC)

Fig. 3. Varying Budget in NYC (a, b), LA (c, d), Varying θ in NYC (e, f), Varying No. of Slot in NYC (g), Varying ϵ in NYC (h, i)

zones. Third, in the case of the NYC dataset varying trajectory size from $40k$ to $120k$, the computational time increases drastically in BBS compared to the BFBS and Greedy approach. The BBS and Greedy outperform the BFBS and baseline methods by up to 40% to 60% in terms of influence with the increase in trajectory size.

5 Conclusion and Future Research Direction

In this paper, we study the problem of influential billboard slot selection under zonal influence constraint and formulate this as a discrete optimization problem. We propose a simple greedy and branch and bound approach with two bound estimation techniques to address this problem. We experimented with two real-world billboards and trajectory datasets and analyzed their time and space requirements. We also analyzed proposed methods to obtain the performance guarantee. Our proposed BBS gives an optimal solution; however, its computational time requirements are its bottleneck. Therefore, our future work on this problem is to develop more efficient solution methodologies.

References

1. Ali, D., Banerjee, S., Prasad, Y.: Influential billboard slot selection using pruned submodularity graph. In: Chen, W., Yao, L., Cai, T., Pan, S., Shen, T., Li, X. (eds.) Advanced Data Mining and Applications, pp. 216–230 (2022)
2. Ali, D., Banerjee, S., Prasad, Y.: Influential billboard slot selection using spatial clustering and pruned submodularity graph. arXiv preprint arXiv:2305.08949 (2023)
3. Ali, D., Banerjee, S., Prasad, Y.: Minimizing regret in billboard advertisement under zonal influence constraint. arXiv preprint arXiv:2402.01294 (2024)
4. Ali, D., Banerjee, S., Prasad, Y.: Regret minimization in billboard advertisement under zonal influence constraint. In: Proceedings of the 39th ACM/SIGAPP Symposium on Applied Computing, pp. 329–336 (2024)
5. Ali, D., Bhagat, A.K., Banerjee, S., Prasad, Y.: Efficient algorithms for regret minimization in billboard advertisement (student abstract). In: Proceedings of the AAAI Conference on Artificial Intelligence, vol. 37, no. 13, pp. 16148–16149 (2023)
6. Wang, L., et al.: Data-driven targeted advertising recommendation system for outdoor billboard. ACM Trans. Intell. Syst. Technol. (TIST) **13**(2), 1–23 (2022)
7. Wang, L., Yu, Z., Yang, D., Ma, H., Sheng, H.: Efficiently targeted billboard advertising using crowdsensing vehicle trajectory data. IEEE Trans. Ind. Inf. **16**(2), 1058–1066 (2020)
8. Zahrádka, J., Machová, V., Kučera, J.: What is the price of outdoor advertising: a case study of the Czech republic? Ad Alta: J. Interdisc. Res. (2021)
9. Zhang, P., Bao, Z., Li, Y., Li, G., Zhang, Y., Peng, Z.: Trajectory-driven influential billboard placement. In: Proceedings of the 24th ACM SIGKDD International Conference on Knowledge Discovery & Data Mining, KDD 2018, pp. 2748–2757. Association for Computing Machinery, New York (2018)
10. Zhang, P., Bao, Z., Li, Y., Li, G., Zhang, Y., Peng, Z.: Towards an optimal outdoor advertising placement: when a budget constraint meets moving trajectories. ACM Trans. Knowl. Discov. Data (TKDD) **14**(5), 1–32 (2020)
11. Zhang, Y., Li, Y., Bao, Z., Mo, S., Zhang, P.: Optimizing impression counts for outdoor advertising. In: Proceedings of the 25th ACM SIGKDD International Conference on Knowledge Discovery & Data Mining, KDD 2019, pp. 1205–1215. Association for Computing Machinery, New York (2019)
12. Zhang, Y., Li, Y., Bao, Z., Zheng, B., Jagadish, H.V.: Minimizing the regret of an influence provider. In: Proceedings of the 2021 International Conference on Management of Data, SIGMOD 2021, pp. 2115–2127. Association for Computing Machinery, New York (2021)

Amethyst - A Generalized on-the-Fly De/Re-compression Framework to Accelerate Data-Intensive Integer Operations on GPUs

Johannes Fett(✉), Dirk Habich, and Wolfgang Lehner

Database Research Group, Technische Universität Dresden, Dresden, Germany
{johannes.fett,dirk.habich,wolfgang.lehner}@tu-dresden.de

Abstract. In this paper, we present *Amethyst*, a generalized on-the-fly de/re-compression framework for GPUs. Our developed framework allows to execute every computational function on every (un)compressed input format and to output the result in an arbitrary (un)compressed format. Thus, our on-the-fly de/re-compression framework approach automatically conducts a decompression on the input side and compression on the output side. To show the feasibility and applicability of our framework, we developed a prototype for integer data types using different integer compression formats. We use this prototype to present exhaustive evaluation results using a great variety of data-intensive operations ranging from simple additions up to compaction and compute-intensive matrix multiplications to validate the efficiency of *Amethyst*.

Keywords: GPU · Compression · Null Suppression · Performance Optimization

1 Introduction

In the last decades, graphics processing units (GPUs) have developed into general-purpose computing devices with general-purpose parallel programming models such as CUDA [11] and OpenCL [9]. Due to this evolution, GPUs are increasingly used for high-performance and data-intensive workloads because of their capabilities to accelerate computation using massive parallelism. For example, a modern single GPU can have up to 80 GB of high-bandwidth memory delivering up to 2 TBps of memory bandwidth and 26 Tflops of compute power. However, a fundamental problem for data-intensive workloads on GPUs is that the memory bandwidth is well utilized, while compute utilization is very low. For this, we profiled an element-wise addition of two arrays with integer data on different NVIDIA GPUs. All profiler data shows that the memory throughput of 90% is almost optimal, while only 4–5% of compute resources are used (a representative profiler result is shown Fig. 4 in Sect. 4 with label *uncomp* for an uncompressed binary add ($a + b = x$)).

The reasons for this are manifold. On the one hand, uncompressed data processing is still state-of-the art, as the programming models focus more on efficient parallelization concepts. On the other hand, the conducted computations

J. Tekli et al. (Eds.): ADBIS 2024, LNCS 14918, pp. 107–120, 2024.
https://doi.org/10.1007/978-3-031-70626-4_8

for data-intensive workloads such as the element-wise addition are not compute-intensive per definition. Therefore, the memory bandwidth is always the bottleneck and higher compute resources are not beneficial for data-intensive workloads on uncompressed data. To tackle this issue, compression is a proven technique for accelerating data-intensive workloads by offering the following advantages as demonstrated in [13]: (1) more data can be stored in the main memory, (2) data transfer is optimized as more logical elements are transferred through compression, and (3) compute utilization increases through compression and decompression. However, the following challenges now arise: (i) not every computation can be performed on compressed data and (ii) there is no best compression technique for a specific data type, but this depends, among other things, on the data properties.

Our Contribution: To systematical address these challenges, we present *Amethyst* in this paper. *Amethyst* is a generalized on-the-fly de/re-compression framework for GPUs to accelerate data-intensive workload using compression. Our developed framework allows to execute every computational function on every (un)compressed input format and to output the result in an arbitrary (un)compressed format. Thus, our on-the-fly de/re-compression framework approach automatically conducts a decompression on the input side and compression on the output side, so that only a small part of the data is decompressed for the computation. To show the feasibility and applicability of our framework, we developed a prototype for the integer data type using different integer compression formats. We use this prototype to present exhaustive evaluation results using a great variety of data-intensive operations ranging from simple additions up to complex matrix multiplications to validate the efficiency of *Amethyst*.

Outline: The remainder of the paper is organized as follows: Sect. 2 describes the necessary background and related work for this paper. Afterwards, we introduce our developed on-the-fly de/re-compression framework *Amethyst* including the prototype for integer data in Sect. 3. Then, selective results of our exhaustive evaluation are presented in Sect. 4. Finally, we conclude the paper with a brief summary and an outlook on our ongoing work in Sect. 5.

2 Background and Related Work

In this section, we review the basics of the GPU architecture and describe approaches for data compression on CPU and GPU.

Nvdida GPU Architecture and Execution Model: An Nvidia GPU consists of a set of streaming multiprocessors (SM). Each streaming multiprocessor has its local memory. The register memory and a memory block that is both used for L1 cache and shared memory are physically located on each streaming multiprocessor as exclusive resources. A set of up to 128 arithmetic logic units called CUDA cores is also located on each streaming multiprocessor. L2 Cache is shared between all SMs within the GPU. The largest tier of GPU memory is global memory. Depending on the GPU model, global memory can reach a size of up to 80 GB. Shared memory, on the other hand, only ranges from 48 KiB to 228 KiB. While global memory is larger than shared memory, it has increased latency and reduced bandwidth compared to on-chip memory. Computations on GPUs leverage the *Single Instruction Multiple Thread* programming model. A large set of threads is spawned and performs a computation that is called a kernel. Threads are partitioned evenly into blocks. The sum of all blocks is called a grid. A block is assigned to a streaming multiprocessor. To run a kernel, both the number of blocks and threads per blocks are required parameters. Both parameters together are called a CUDA configuration. Both, memory and compute instructions are performed on groups of 32 threads that are called a warps.

Related Work: The standard procedure for processing data-intensive workloads on GPUs is as follows [12]: In the first step, uncompressed data is copied over from GPU global memory to SM. In the second step, the computation is performed by SMs, while in the third step the result is written back to global memory. A large number of papers have already shown that this standard processing concept can be accelerated by compression on CPU as well as on GPU [4,8,13]. For database systems, as the most representative example for data-intensive workloads, lossless compression techniques are distinguished into *lightweight* and *heavyweight* [4,8]. While lightweight techniques are mainly used for in-memory processing, heavyweight like Huffman or Lempel-Ziv are used to accelerate the transfer from disk to main-memory. The most representative example for lightweight compression of integer data is null suppression (NS) [4,8]. NS removes leading zeros from an integer's bit representation and for this, a large variety of algorithms and formats have been proposed [4,8]. In a recent paper [13], Shanbhag et al. introduced the concept of processing compressed integer data on GPU using a "tile"-at-a-time approach to accelerate query processing. A tile is a chunk of compressed data that fits into shared memory. The tile size is defined by CUDA configuration parameters resulting in a mapping how many elements each thread/block needs to process. The tiles are uncompressed for the computation, but the results are not recompressed for further processing. However, Damme et al. [5] demonstrated that the recompression of results with an appropriate compression technique is beneficial on CPUs to accelerate query processing. With *Amethyst*, we now introduce a generalized framework that enables a continuous compressed processing for arbitrary computations on GPUs.

3 *Amethyst* Framework

As highlighted in Sect. 1, data-intensive GPU kernels may lead to a heavy memory utilization, while compute resources are mostly idle. Our *Amethyst* aims to overcome this shortcoming. While Sect. 3.1 explains our framework approach in general, Sect. 3.2 presents an implemented prototype with a focus on integer data. Then, the operations used to evaluate our prototype are described in Sect. 3.3.

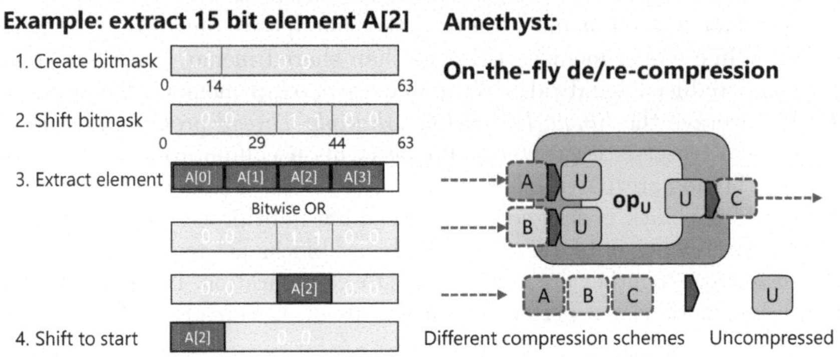

Fig. 1. Amethyst concept overview and example decompression of a 15 bit element.

3.1 General Concept

Figure 1) gives a schematic overview of our framework and key design decisions are:

(i) Data in GPU global memory is always compressed; uncompressed data is never allowed to be stored in global memory.
(ii) On-the-fly decompression and re-compression are first class citizen and happen within GPU kernels that perform operations.
(iii) Any kind of compression scheme can be used.
(iv) Different compression schemes and memory layouts can be combined within one kernel as shown in Fig. 1).
(v) Calculations performed are interchangeable using C++ template programming and language features like function pointers.

In the following, an example for performing an operation with Amethyst is described. After invoking the GPU kernel, compressed input data is copied over from global memory to shared memory. This happens in parallel and distributed across all available streaming multiprocessors. Only the data immediately needed for a calculation is copied. As shown in Fig. 1), data compressed in formats A and B are uncompressed from shared memory into native data types in register space. Then, the chosen operation is performed in register. Afterwards, the result is re-compressed on-the-fly into a different compressed data format C and written back to shared memory. This allows to take growing or shrinking bit sizes depending on the operation into account. Finally, the compressed data is copied back into global memory. A loop repeats this steps to process any size of input data. The computational effort increases due to decompression and re-compression. As profiler data shows, these resources are readily available. Moreover, more logical elements in a compressed form can be transferred from global memory than in an uncompressed in the same time frame. The operation itself is performed on uncompressed native data types. Thus, the calculations are the same as in uncompressed kernels. At no time, uncompressed array data is copied between global memory and shared memory. Compression formats, data types and operations are made flexible by C++ template programming and function pointers. Both can be redefined without changing the invoked GPU kernel.

3.2 Prototypical Implementation for Integer Data

In principle, our framework can be used for all data types and corresponding compression techniques. To demonstrate this, we have implemented an initial prototype for unsigned integer data as heavily used data type in in-memory database systems [5,12,13]. Database systems also heavily use dictionary encoding to store strings as unsigned integers to improve performance [7]. As a consequence unsigned integer processing is of great importance for database systems. For the prototype, we used null suppression (NS) as compression technique, whereby its basic idea is the omission of leading zeros in the bit representation [8]. For the compressed bit representation, we use two different memory layouts: *bit packing* and *bitstream*. While bit packing packs multiple compressed elements into one native, e.g., 64-bit unsigned integer element, bitstream stores compressed elements in a continuous bitstream instead. Based on that, any element bit size between 1 and 64 is supported for both memory layouts.

An example how to decompress a single element from a bit packed data format is shown in Fig. 1. A bitmask of the same size as the compressed element is created within a 64-bit element in step 1. A shift moves the bitmask to the same bit index as the compressed element (step 2). In step 3, a bitwise OR is used to select the compressed element from a bit-packed 64-bit element. Finally, a shift is used to shift the data element to bit index 0 (step 4). This results in a native uncompressed data element in register memory space. We implemented compression as well as decompression routines for NS with the two different memory layout within GPU kernels. For each memory layout, a different kernel is provided. The memory layout like bit packing is static within the kernel.

Listing 1.1 shows how to configure a binary operation based on NS bit packing. The input bit size, the output bit size, and the performed operation are flexible and defined via template functions at compile time. The GPU bit packing binary operation kernel provides all code for managing memory transfers, decompression, re-compression, and writing back to global memory. Compressed data in NS bit packing format of a flexible bit size is copied from global memory to shared memory. Elements are unpacked from shared memory into native data types in register space. The kernel needs to be combined with additionally defined functions. This decouples compression and computations performed on the data. These functions shown in Listing 1.1 provide two key pieces. First, the operation that the kernel performs is defined as template function as shown in lines 2 to 4 for a simple addition. The function takes two native data types as input parameters and returns a native data type as well. In this example an addition of two elements is implemented. A bitwise AND with a bitmask is applied to ensure that the outgoing element does not exceed the bit size defined by the template parameter. Secondly, the output bit size is defined in another template function. For add the bit size function is shown in lines 16 to 18. Finally, both the operation and the output bit size calculation are combined in lines 25 to 27 into the BINARY_OP function. The GPU kernel then calls the BINARY_OP function. This allows redefining the performed operation, as well as the output bit size outside the GPU kernel definition. Alternative operations and bit size calculations are also shown in Listing 1.1. These can be plugged into the kernel instead of add. Defining new functions in this manner is also possible and shows the flexibility of *Amethyst*.

```
1 //example operation 1: add (a + b = c)
2 template <int bits>
3 constexpr __host__ __device__ uint64_t op_add(uint64_t l,
    uint64_t r)
4 {   return (l + r) & gen_bitmask(bits);          }
5
6 //example operation 2: a more complex math formula
7 template <int bits>
8 constexpr __host__ __device__ uint64_t op_math(uint64_t l,
    uint64_t r)
9 {   return ((l*l + r*r + 4*l) / (l*l + r*r +5*l)) &
    gen_bitmask(bits); }
10
11 //example operation 3: boolean logic
12 template <int bits>
13 constexpr __host__ __device__ uint64_t op_bool_complex(
    uint64_t l, uint64_t r)
14 {   return ((l && !r) || (!l && r)) & gen_bitmask(bits);  }
15
16 //defining output bit size for add:
17 constexpr __host__ __device__ int op_add_outbits(int bits)
18 {  return bits == 64 ? 64 : bits + 1;          }
19
20 //defining output bit size for boolean logic:
21 constexpr __host__ __device__ int op_bool_outbits(int bits)
22 {   return  1;                                  }
23
24 //Combining operation and output bit size for GPU kernel
25 template <int bits>
26 constexpr __host__ __device__ uint64_t BINARY_OP(uint64_t a,
    uint64_t b)
27 {   return op_add<op_add_outbits(bits)>(a, b);     }
```

Listing 1.1. Defining different operations and compressed formats for binary element-wise operations.

3.3 Amethyst Prototype Algorithm Overview

To showcase the feasibility and limits of our *Amethyst* prototype, we implemented a variety of different operations. An overview is shown in Table 1 and explained in the following. **Binary Operation** is any element-wise operation

Table 1. An overview of all available *Amethyst* algorithms

Algorithm	Input	Output	Flexible OP
Binary Operation	2 arrays	1 array	yes
Horizontal Operation	1 array	1 scalar	yes
Filter	1 array, 1 predicate	1 bitmask array	no
Compaction	1 array, 1 bitmask array	1 array	no
GroupBy	2 arrays	2 arrays	yes
Matrix multiplication	2 arrays	1 array	no

on two input elements resulting in one output element. It iterates over two input arrays and writes back into one output array. It supports any operation on two `uint64` elements that compiles with C++ 17 compiler. Examples of this are shown in Listing 1.1. As output bit size depends on operation and input bit size, it needs to be defined as function. **Horizontal Operation** is performed across all elements of an input array and outputs a single scalar value. The operation is flexible as well. Examples are sum of all elements, count of all elements, Boolean logic or bitwise logic. **Filter operation** compares the values of an input array against a predicate. The output is a bitmask array. For reach comparison a 1 is written back if the predicate is true, or a 0 if it is false. **GroupBy** is a common database operator [2]. The input consists of an array of keys and values. Grouped by keys, a horizontal operation is applied on all values. GroupBy performs a horizontal operation across partitions of values with the same key. An example is the revenue of sales per country if values are individual sales and keys are countries. GroupBy usually uses a hash-table based approach and atomic operations. The hash-table resides completely in shared memory instead of global memory. The larger the group count, the larger the hash table grows. For hash table size, the worst case of group count equal to element count is assumed. Using compression, the hash table sizes fit into shared memory that can not be processed on GPU shared memory otherwise. **Compaction** is an operation that selects a subset of elements by the means of a bitmask and writes them back contiguously. The input is an array of data elements and a corresponding bitmask, where a bit set to 1 selects elements. It is difficult to pallelize compaction, as the memory offset of each output element depends on the previously selected elements. Calculating the memory offsets is the first step before any data is written into the output array. The input data is partitioned into chunks of 1024 elements. A popcount is applied to each chunk to calculate the total memory offset of each single chunk. Additionally within each chunk a prefix sum is calculated to calculate the memory offset per element within a chunk. The final memory offset per element is the sum of the internal memory offset of a chunk and the popcounts (number of selected elements) of all previous chunks. In [6], we introduced an optimization to the state-of-the-art GPU NVIDIA compaction algorithm CUB::deviceselect [3].

By introducing a skipping mechanism for empty chunks, memory accesses can be avoided which improves overall performance in most cases. *Amethyst* improves the SPACE compaction algorithm by combining the skipping optimization with data compression. **Matrix multiplication** is a compression based unsigned integer matrix multiplication. It leverages shared memory and a tile based approach based on NVIDIA documentation [10]. Instead of loading a row-at-a-time, a tile is loaded into shared memory. A tile consists of a submatrix. Two tiles can be used to calculate a partial result of one output submatrix. The partial results are stored in shared memory in a bitpacked compressed format. Results of additional tiles are added to the partial results in a loop until the final result is calculated. As bit size can grow significantly for matrix multiplication, Amethyst offers different strategies to deal with output bit sizes and overflows. This allows to exploit statistical knowledge about input matrix values. Offered output bit size strategies are: same as input (sai), ceil to next power of two equal or greater than input bits (c2p2), maximum largest possible output value (maxv). In case a result does not fit into the output bit size two resolution strategies are offered: overflow (ovf) and saturate (sat). Overflow is the default GPU device behavior and will cycle through the value space. Saturate pins the value at maximum if an overflow occurs. Both, bitstream and bit packing formats are supported.

4 Evaluation

All *Amethyst* algorithms are implemented in C++ 17 using CUDA 12.3 and available open-source [1]. CMake 3.28.3 is used as build system. The GPU used for all experiments is a NVIDIA RTX 8000 Quadro. *Amethyst* compiles both with the NVIDIA-supplied proprietary nvcc compiler and the open source Clang compiler. CUDA versions 10.x and lower are unable to compile *Amethyst*, due to lacking C++ 17 support. The total runtime of each experiment is measured using CUDAevents. Uncompressed inputs are 1 GiB uint64 elements per array. For example, the uncompressed element-wise addition using the binary operation kernel would use 2 GiB (2 arrays * 1 GiB) input data size and 1 GiB output data size. All *Amethyst* algorithms are evaluated across a range of bit sizes, CUDA configurations and operations. Speedup is calculated as follows $\frac{compressed\_runtime}{baseline\_runtime}$. The run-time used is the average of two runs per parameter set. The lowest run time across all CUDA configurations per bit size is used in the evaluation. The fastest baseline kernel run across all bit size and CUDA configurations is shown. An exception to this is the GroupBy kernel. For GroupBy, a synthetic baseline was calculated. As a simple element-wise add reaches 93% memory bandwidth, a more complex GroupBy is unlikely to match that. 93% memory bandwidth for processing the 2 GiB of input data is chosen as a pessimistic baseline for a lack of a reference GroupBy implementation. The reference baseline for matrix multiplication is an NVIDIA supplied optimized matrix multiplication guide [10]. It works only on uncompressed data and follows the output scheme same as input (sai). A modified version that checks for overflows and pins the output to maximum value is used as baseline for all

Amethyst matrix multiplications with saturate overflow resolution strategy. For compaction, the Nvidia supplied solution `cub::DeviceSelect:Flagged` [3] is chosen as baseline. Compaction is heavily influenced by the bitmask that selects the written out data array. Both, the percentage of selected data and how the 1 s are distributed across the bitmask influence the results. Three different bit mask distributions and different percentages of selected data have been generated for the evaluation. Single cluster consists of one block of ones, while multi cluster partitions the ones into 4 equidistant blocks across the bitmask. Uniform distribution uniformly distributes all ones across the bitmask array.

Evaluation Results: Figure 2 shows experiments with all *Amethyst* algorithms except matrix multiplication. Figure 3 shows matrix multiplication experiments. On the Y-axis, the speedup is shown. The X-axis shows input bit sizes per element ranging from 64 to 1 bit. A grey dotted baseline at $Y = 1$ shows the uncompressed baseline approach. Continuous lines represent bitstream compressed formats, while dashed lines show bit packing based implementations. **Binary Operation:** The top 3 graphs show different calculations and bit sizes performed on the binary operation kernel. Math operations contain a number of arithmetical operations. The series named Div, mul, mod, and add are basic arithmetical operations. The math series calculates the following formula: $\frac{(l*l+r*r+4*l)}{(l*l+r*r+5*l)}$ as shown in Listing 1.1 line 14. For bit sizes lower than 14, Bit packing outperforms the bitstream approach. For 1 bit element size, add achieves a speedup of 18.76 with bit packing versus a bitstream speedup of 10.8. Between 33 and 64 bit, bit packing approaches can only contain 1 element and have no memory size advantage over uncompressed data. However, the performance loss against uncompressed baseline is minimal with a speedup of 0.999 for add. Bit packed div has the worst case performance loss with a speedup of 0.9982 against baseline. For math series there are two data points where the speedup is 1.00003 without compression. Using compression based algorithms has negligible performance impact even when no compression occurs. In all series the highest speedup across all bit sizes is achieved in bit-packing format. The bitstream approach results in performance gains starting from a bit size of 63 for add and mul. Mod, div, and math are the worst performing series. For a bit size of 44 and lower, bitstream outperforms the uncompressed baseline. The second graph *Bitwise Operations* shows bitwise arithmetic operations. bitwise_simple is a bitwise OR while bitwise_complex is $(l \wedge \neg r) \vee (\neg l \wedge r)$. While both series show similar behavior, the simple operation always outperforms the complex one. The best case speedups are 19.22 for the simple operation at 1 bit and 17.43 for the complex at 2 bit. The worst case speedup is 0.997. *boolean operation* applies the previously mentioned bitwise logic operations instead on an element level as boolean logic operations. Thus the resulting output bit size becomes 1. For C++, any boolean operation can be performed on integer data types. In that case, any value greater than 0 equals true. While this appears to be a rather niche operation, it shows the flexibility of *Amethyst.* **Horizontal Operation:** The Graph *Horizontal operation* shows an operation across all elements of an array like a sum of all elements. The graph behaves similar to Binary operation

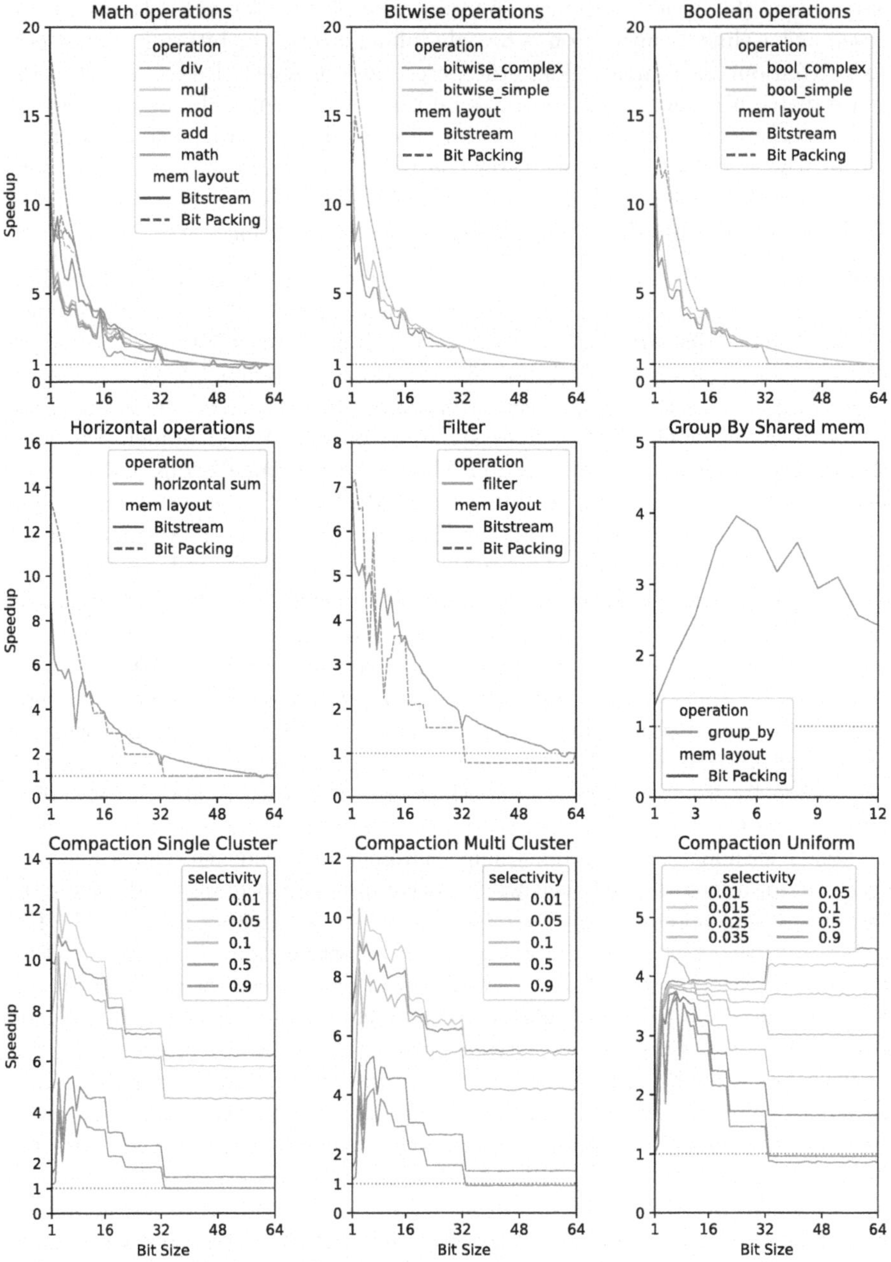

Fig. 2. Evaluation of compressed *Amethyst* algorithms against an uncompressed baseline approach (grey dotted line at y=1). A variety of different operations is used on the binary operation kernel without changing GPU kernel code.

but has slightly lower speedup. Best case speedup is 13.41 at 1 bit. **Filter Operation:** This operation shows consistent speedups for both memory layouts. The highest speedup 7.23 is achieved with bitstream layout at bit size 1. **GroupBy Shared Memory:** This is only done on bit size between 1 and 12 bit. Greater bit size would not fit into shared memory, as the hash table needs to fit into shared memory. Compression allows to perform shared memory based GroupBy with a group count, which is not possible to process without compression on GPU. It achieves the best case speedup at 5 bits with 3.96. **Compaction:** For Compaction only a bit packing layout has been implemented. As expected, no performance difference occurs between 32 and 64 bit size. While Multi and single cluster bitmask distributions lead to similar results, there is a notable difference to uniform bitmask distribution. Single and multi cluster distributions increasingly outperform the baseline below 32 bit. Between 22 and 31 the speedup only varies between 7.05 and 7.12. The reason is that the number of compressed elements that can be packed into a 64 bit element is 3 within this bit sizes' interval. The best case speedup depends on the % of selected data (selectivity * 100). In case of 90% single cluster, the best case is a speedup of 4.22 at 7 bits. For 10% the best case speedup is 10.30 at 3 bits. The major difference to uniform is the number of chunks that can be skipped. Single and multi cluster allow skipping of most chunks as the ones are tightly packed into consecutive slices (4 for multi cluster and 1 for single cluster). In case of using uniform distribution with very low % of selected data, reaching 32 bit actually reduces the performance. At % 1 selected data, most chunks are skipped. Scattered elements are read and written back. The extra control flow for compression outweighs the compression benefit if almost no elements are written back. **Matrix:** Each graph represents a combination of two parameters: The chosen output bit size scheme and how overflows are handled. An explanation of these parameters is provided in Sect. 3.3. Default GPU device overflow behavior is cycling through the value space (ovf) after exceeding the maximum value (shown at the top row). In case of ovf as overflow strategy, increasing speedups are achieved at 32 bit size and lower. C2p2 and sai reach peak speedups 4.99 and 4.15 at 8 and 11 bit size. Maxv_ovf has the speedup at bit size 1 unlike c2p2_ovf and sai_ovf. For the overflow resolution strategy saturate (sat), the speedup is overall far below ovf strategy. The extra computation and branching for overflow handling further reduces the benefit of compression. As matrix multiplication makes heavy use of intermediate data in shared memory, the global memory layout has far lower impact compared to the experiments shown in Fig. 2. Overall, there is no one size fits all approach. Both, bit packing and bitstream are needed to reach good performance across all compressed bit sizes. For lower bit sizes, with the exception of matrix multiplication, bit packing increasingly outperforms bitstream approaches, while bitstream starts to offer performance benefits as early as 1 compressed bit. In most cases significant speedups can be achieved. On-the-fly re-compression allows the output data of kernels to be used directly as input data for additional kernels. Generally, the lower the bit size the higher the speedup. With the exception of bit sizes 1 to 4, complex operations like GroupBy and

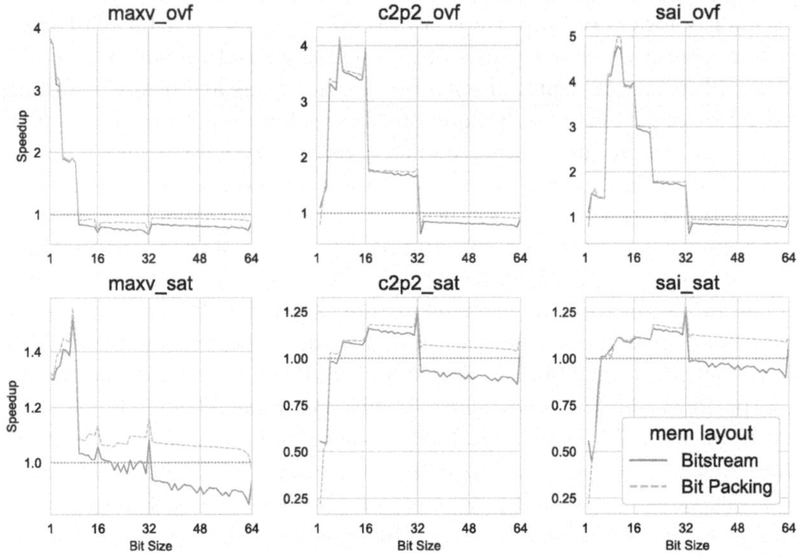

Fig. 3. Evaluation of compressed *Amethyst* matrix multiplication against an uncompressed baseline approach (grey dotted line at y=1). A variety of different strategies for handling overflows and output bit size is shown.

compaction become slower for very low bit sizes. A simple add reaches its best case speedup at a bit size of 1. Compaction and GroupBy both perform multiple operations and have a complex control flow. In extreme cases the extra instruction needed to perform the on-the-fly de-compression and re-compression leads to compute bottlenecks. While the low bit sizes for complex operations perform worse than slightly higher bit sizes, they still outperform the uncompressed algorithms. Using multiple and expensive operations for the element-wise binary kernel lead to lower speedup than simple operations overall, but still outperforms baseline. Matrix multiplication can also benefit from compressed data formats both in global memory and shared memory. However, the speedup is much higher for data intense operations. The global memory bottleneck is the major obstacle to improve data processing performance on GPU. *Amethyst* shows that a variety of different operations can be significantly sped up by using compression.

Profiler Results: In addition to the speedup results, a selection of different compressed bit sizes have also been profiled using the Nsight compute profiler on an element-wise binary addition. For comparison, the same operation on uncompressed data reaches 93.01% memory throughput and 4.51% compute throughput. For the compressed kernel, the hardware utilization improves. At a bit size of 8, memory throughput is 91.81% and compute throughput is 51.19%. Lower bit sizes lead to decreased memory throughput, as the previous memory-bottleneck increasingly shifts to a compute bottleneck. For the simple add operation with bit packing the sweet spot for best overall hardware utilization is at bit size 5 with 87.82% memory throughput and 73.8% compute throughput. Bit sizes that have both high memory and compute throughput are closest to optimal

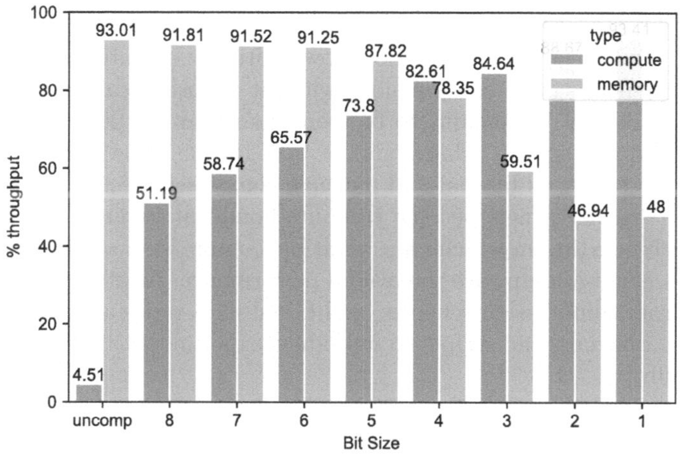

Fig. 4. NVIDIA Nsight profiler data about GPU memory and compute throughput for compressed binary add (a + b = c). bit packing memory layout. Ranging from 8 to 1 bit compressed bit sizes and 64 bit uncompressed

speedup. A bit size of 5 has a compression ratio of 12 and a speedup of 11.1. For lower bit sizes the gap between compression ratio and speedup increases. At 1 bit the compression ratio is 64, but the speedup is only 18.76. Overall, hardware utilization improves as shown in Fig. 4.

5 Conclusion and Future Work

Amethyst enables major performance gains for memory-bound operations to accelerate data-intensive workloads. The concept of On-the-fly de-/re-compression overall works very well on GPUs due to an abundance of compute resources. Flexible operations via C++ template programming can be used to combine standard C++ functions on native data types with CUDA based GPU compression kernels. Both, inputs and outputs are always compressed and can be reused without invoking extra compression kernels. This is a novel approach for GPU data compression and advances the state-of-the-art. The compute utilization steadily increases when bit sizes decrease. At very small bit sizes, the kernels become compute-bound instead of memory-bound. *Amethyst* kernels reach a far better overall hardware utilization compared to uncompressed kernels. Considering compression memory layouts, there is no one size fits all approach [4]. Above 32 bit size, a bitstream based approach leads to performance gains, whereas bit packing has no advantage. At very low bit sizes, bit packing generally outperforms bitstream based approaches. Combining the low shared memory footprint of *Amethyst* with compressing intermediate data in shared memory as well, allows fitting bigger problem sizes into shared memory.

As a future work, we want to explore abstracting the on-the-fly de-/re-compression technique into an array like memory access. Instead of tailoring highly specific compressed memory operations on a kernel basis, this is done once on a memory level. This would allow writing data-intensive kernel code that highly benefits from compression, without tailoring compressed memory access to each kernel. It remains to be seen, if such an abstraction concept for on-the-fly compression can achieve similar speedups as shown in the *Amethyst* prototype. Moreover, additional and multiple compression schemes are of interest. As even simple kernels become compute-bound at lower bit sizes, it is an open research question in which cases multiple compression schemes for on-the-fly de-/re-compression improve the overall performance. As all computations are decoupled from compression schemes, additional data types are also of interest. Source code, raw data and scripts to reproduce experiments are provided on our public Github [1].

Acknowledgements. This work was partly funded by (1) the European Union's Horizon 2020 research and innovation program under grant agreement No 957407 (DAPHNE) and (2) SoftwareCampus program of the German Federal Ministry of Education and Research.

References

1. Amethyst Github. https://github.com/yogi-tud/Amethyst
2. Chaudhuri, S., Shim, K.: Including group-by in query optimization. In: VLDB, vol. 94, pp. 12–15 (1994)
3. CUB: Main Page. https://nvlabs.github.io/cub/index.html
4. Damme, P., Ungethüm, A., Hildebrandt, J., Habich, D., Lehner, W.: From a comprehensive experimental survey to a cost-based selection strategy for lightweight integer compression algorithms. ACM Trans. Database Syst. **44**(3), 9:1–9:46 (2019)
5. Damme, P., Ungethüm, A., Pietrzyk, J., Krause, A., Habich, D., Lehner, W.: Morphstore: analytical query engine with a holistic compression-enabled processing model. Proc. VLDB Endow. **13**(11), 2396–2410 (2020)
6. Fett, J., Kober, U., Schwarz, C., Habich, D., Lehner, W.: Accelerating parallel operation for compacting selected elements on GPUs. In: Cano, J., Trinder, P. (eds.) Euro-Par 2022. LNCS, vol. 13440, pp. 186–200. Springer, Cham (2022). https://doi.org/10.1007/978-3-031-12597-3_12
7. Graefe, G., Shapiro, L.D.: Data compression and database performance. University of Colorado, Boulder, Department of Computer Science (1990)
8. Lemire, D., Boytsov, L.: Decoding billions of integers per second through vectorization. Softw. Pract. Experience **45**(1), 1–29 (2015)
9. Munshi, A., Gaster, B., Mattson, T.G., Ginsburg, D.: OpenCL Programming Guide. Pearson Education (2011)
10. NVIDIA Blog: Write Flexible Kernels with Grid-Stride Loops. https://docs.nvidia.com/cuda/pdf/CUDA_C_Best_Practices_Guide.pdf
11. Sanders, J., Kandrot, E.: CUDA by Example: An Introduction to General-Purpose GPU Programming. Addison-Wesley Professional (2010)
12. Shanbhag, A., Madden, S., Yu, X.: A study of the fundamental performance characteristics of GPUs and CPUs for database analytics. In: SIGMOD, pp. 1617–1632 (2020)
13. Shanbhag, A., Yogatama, B.W., Yu, X., Madden, S.: Tile-based lightweight integer compression in GPU. In: SIGMOD, pp. 1390–1403 (2022)

Access Methods and Query Processing

GXJoin: Generalized Cell Transformations for Explainable Joinability

Soroush Omidvartehrani(✉), Arash Dargahi Nobari, and Davood Rafiei

University of Alberta, Edmonton, Canada
{s.omidvartehrani,dargahi,drafiei}@ualberta.ca

Abstract. Describing real-world entities can vary across different sources, posing a challenge when integrating or exchanging data. We study the problem of joinability under syntactic transformations, where two columns are not equi-joinable but can become equi-joinable after some transformations. Discovering those transformations is a challenge because of the large space of possible candidates, which grows with the input length and the number of rows. Our focus is on the generality of transformations, aiming to make the relevant models applicable across various instances and domains. We explore a few generalization techniques, emphasizing those that yield transformations covering a larger number of rows and are often easier to explain. Through extensive evaluation on two real-world datasets and employing diverse metrics for measuring the coverage and simplicity of the transformations, our approach demonstrates superior performance over state-of-the-art approaches by generating fewer, simpler and hence more explainable transformations as well as improving the join performance.

Keywords: Generalization · Explainable Join · Transformation

1 Introduction

Generalization is a fundamental concept across various branches of science. In the realm of machine learning, the notion of model generalizability serves as a measure of how effectively a model, learned from a limited set of training samples, can extend its performance to unseen data during testing. We study the problem of generalizability in transforming tables for joins, where the task is to transform data from one table formatting to another to make them equi-joinable, using only a limited sample of joinable row pairs. In this context, the transformations are used to explain and validate the join process and the generalizability of the transformations signifies the capacity to accurately represent the true mapping rather than capturing some form of *accidental regularity* in the data. Also, a more general transformation will garner more supporting evidence from the data and may be deemed more reliable, as opposed to a transformation that may capture random noise in the data. The problem of transforming tables has been explored in the literature [8,9,18,30], with some variations in the supported set of transformations. However, the generality aspects of these transformations have not been investigated in these studies.

J. Tekli et al. (Eds.): ADBIS 2024, LNCS 14918, pp. 123–137, 2024.
https://doi.org/10.1007/978-3-031-70626-4_9

(a) Course Catalog

Course	Title	Instructor
CMPUT 291	Intro to Databases	Davood Rafiei
CMPUT 391	Database Systems	Mario Antonio Nascimento
CMPUT 175	Intro to Computing	Osmar Zaiane

(b) Linkedin

Linkedin Profile	Contact Email
/davood-rafiei-a215b65	drafiei@ualberta.ca
/mario-nascimento-a9362a3	mnascime@ualberta.ca
/zaiane	zaiane@ualberta.ca

(c) Publications

Author	Publication
D. Rafiei	Similarity-based queries for time series data
M. A. Nascimento	On the generation of spatiotemporal datasets
O. Zaiane	Building a recommender agent for e-learning systems

Transformations from CST:

(a): source → (b): target

T_1: <substr(0, 1), splitSubstr(' ', 1, 0, 6), literal('@ualberta.ca)>

T_2: <substr(0, 1), splitSubstr(' ', 2 , 0, 7), literal('@ualberta.ca)>

T_3: <splitSubstr(' ', 1, 0, 6), literal('@ualberta.ca)>

(a): source → (c): target

T_1: <substr(0, 1), literal ('. '), split(' ', 1)>

T_2: <substr(0, 1), literal ('. '), splitSubstr(' ', 1, 0, 1), literal ('. '), split(' ', 2)>

General Transformations:

T_G: <substr(s, s+1)$^?$, splitSubstr(' ', e, s, s+7), literal('@ualberta.ca)>

T_H: <(splitSubstr(' ', s, s, s+1), literal ('. '))*, split(' ', e)>

Fig. 1. Example of joinable tables and transformations

Motivating Example: A data analyst is tasked to integrate information about faculty members from various sources to create comprehensive profiles. Consider the tables presented in Fig. 1 from three different sources. Despite describing the same entities, these tables are not equi-joinable due to formatting mismatches. The manual construction of a mapping between these tables can be both time-consuming and may necessitate domain knowledge that the analyst might not possess. Suppose the analyst aims to map the instructor name in Table (a) to the contact email in (b) and the author name in (c). Leveraging a recent work on transforming tables [18], referred to as CST and regarded as state-of-the-art for its improvement upon previous works, the analyst obtains three transformations, denoted as T_1, T_2, and T_3 to transform the instructor name in Table (a) to email address in (b), as shown in the Figure. T_1 caters to the first row, extracting the first character of the first name, concatenating it with the last name, and adding a literal to generate the email address in the target table. T_2 follows a similar approach, but accommodates the second row, which includes a middle name. T_3 constructs the email based on the last name only and is applicable to the third row. However, with each transformation applicable to only a single row, the generality of them can be questioned. On the other hand, the transformation T_G, which utilizes optional fields and relative indexes, works universally for all three rows, offering a higher level of generality. One could argue that T_G better captures the underlying pattern and is more readable. The same rationale applies to the mapping of (a) to (c), where CST identifies two transformations to map the tables, whereas a more general transformation, denoted as T_H, exists, covering all table rows[1].

[1] The transformation functions and generalization methods are further explained in Sect. 4.

Problem Studied: We prefer more general transformations that cover a larger number of input rows, as they are more likely to cover unseen data, may better show the underlying patterns in the data, and make it easier for a data scientist to validate the join process. However, there has been limited exploration into which principles of generality can be applied to transformations and whether these generalizations are effective when applied to tabular data. We study this problem in the context of transforming tables for explainable joinability, with the joined columns typically describing the same entities.

Our Approach: Our work aligns with the line of work on program synthesis, aiming to discover cell transformations that enable an equality join. It also falls within the realm of *explainable join*, where transformations explain the reasons behind a join [18,30], albeit the discovered transformations may also aid in predicting missing values [8,22], conducting schema mapping [2,4], and performing data cleaning and repairing [10]. Our work explores several generalization concepts in transforming tables, incorporating those observed in our motivating example, and explores their impact on reducing the number of transformations while maintaining or improving their coverage. Our evaluation on two real-world datasets obtained from different sources reveals that generalization techniques significantly enhance transformation coverage, reduce the number of transformations required to map all rows, better address unseen data, and enhance the end-to-end result of the joinability task. Our contributions can be summarized as follows: (1) We explore the impact of generalization on transforming tables for explainable joinability, aiming to improve the adaptability and coverage of transformations. (2) We expand the domain of transformations by incorporating generalization concepts, including length invariance, repeating patterns, and bi-directionality. (3) Our comprehensive experiments unveil the superior performance of our method, compared to a state-of-the-art approach, across various metrics, including the coverage of transformations, the number of required transformations for mapping, and the ability to cover previously unseen data.[2]

2 Related Works

The pertinent works related to ours can be categorized into (1) finding related tables and joinable row pairs, (2) discovering example-driven table transformations to enable an equi-join, and (3) model generalization.

(1) Finding Related Tables and Joinable Row Pairs: When data to be integrated are located in a data lake, a common initial step involves identifying related tables and potential joinable columns. Finding related tables is studied within the context of joinable tables [3,29,31], unionable tables [15,17], and tables that are semantically similar [26,27]. With relevant tables retrieved, many approaches, including ours, rely on a set of provided examples–whether human-provided or automatically generated–to perform a tabular join through some transformations. The problem of finding joinable row pairs where the matching

[2] Our code and the datasets are available at github.com/soroushomidvar/GXJoin.

pairs are not equi-joinable is well-studied [5,24,25,28], with methods ranging from basic textual similarity functions [5] to bipartite graphs of tokens and their similarity [24,25], and deep learning-based entity matching techniques [14,28]. While the predominant focus of this line of work revolves around matching relevant rows, our approach is centered on the generalization of explainable transformations that render two rows equi-joinable. Hence these approaches can serve as preprocessors in our approach, aiding in the generation of examples.

(2) Discovering Example-Driven Table Transformations: The transformation of cells within a table is recognized as a crucial operation in tabular data integration [11,19], and the discovery of such transformations, guided by a set of examples, has been the subject of several studies [8,9,18,22,30]. FlashFill [8,9] and BlinkFill [22], the two pioneering contributions in this area, create an input graph based on user-provided examples and traverse the graph to generate transformations. Nevertheless, they heavily rely on accurate input examples and struggle when the examples are noisy. The huge search space and resource-intensive computation in FlashFill have inspired the subsequent works to prioritize a smaller search space over generalizability. For instance, Auto-join [30] employ predefined string-based units to generate transformations and a recursive backtracking algorithm to find an optimal one. Limiting the search space to a predefined set of transformations allows Auto-join to handle minor input noise. In a more recent study, Common String-based Transformer (CST) [18] constrains the search space by utilizing common text sequences between source and target examples as textual evidence to construct the skeleton of transformations. The primary focus of all the aforementioned approaches has been on identifying transformations that cover a given example set, with less focus on the generalizability of the transformations to newly added data and unseen examples.

(3) Model Generalization: Within the broader context of our work, there exists a substantial body of literature on generalizing models for diverse inputs. Improving the generalization capacity of neural architectures, often confined to particular domains and tasks, is a well-studied topic [6,12,16,20]. In a closely aligned work, Shi et al. [21] study the compositional generalizability of transformer models, where they identify high-level categories of generalization, such as length generalization and mixing and matching of artifacts to improve the capacity of transformer models trained on simple subtasks to solve more complex tasks. Moreover, There is also a growing interest in meta-learning, enabling models to adapt to new tasks or domains by leveraging knowledge acquired from previous tasks. Some recent works [7,13,23] identify abstract features that prove useful across different problem domains, hence enhancing generalization capabilities. To the best of our knowledge, the generalization of cell transformations for explainable joinability and its impact on performance have not been previously explored.

3 Preliminaries

In this section, we discuss the necessary foundation for our framework and the methods that our approach builds upon.

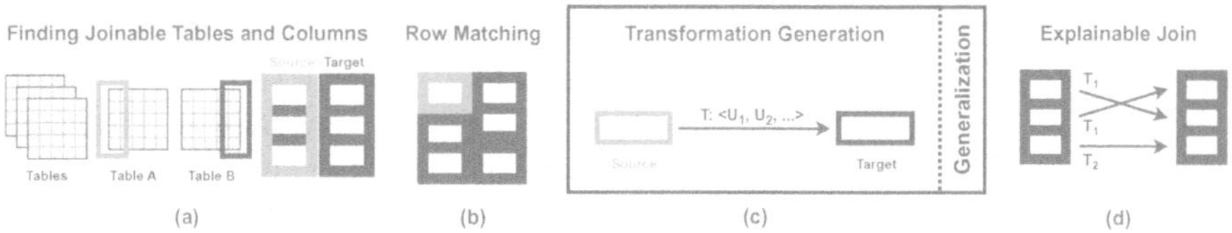

Fig. 2. The end-to-end process of an explainable join

3.1 An End-to-End Explainable Join Algorithm

Figure 2 depicts an overview of a typical unequal join framework. The process begins by identifying relevant tables and joinable columns, followed by the formation of source-target pairs for potentially joinable rows. In some cases, data scientists may provide example row pairs. String-based example-driven transformers such as CST [18] utilize the provided example set to construct transformation skeletons and generate a large set of candidate transformations along with their coverage. With candidate transformations and their coverage known, a transformation may be selected based on its coverage, or a minimal set of transformations covering all given example row pairs may be sought. Our work is on the generalization of the framework, where some strategies are applied either during or after the transformation generation phase, evolving them into more general ones, which can lead to better coverage and a reduced set of covering transformations.

3.2 Cell Transformations

A *cell transformation* (in short, a transformation) is a function that takes an input cell value represented in the source formatting and generates the representation of that value in a desired target formatting. We follow a recent line of work relying on cell transformations to enable unequal joins [18,30] where cell transformations are constructed as a sequence of basic operations referred to as *units*. Each unit is applied on a cell value in the source and generates a fraction of a value in the target. The outputs of multiple units may then be concatenated to form the full cell value in the target. The units are string-based functions that either copy part of the input or a constant literal to the output. A wide range of functions have been employed as units in the literature [9,18,22,30] and our work adopts four commonly occurring units from previous studies [18,30]:

- **literal**(str) returns str as the output,
- **substr** (i, j) provides the substring starting at index i and ending at index j or the end of the input if the input length is shorter,
- **split**(c, t) tokenizes the input based on character c and returns the tth token,
- **splitSubstr**(c, t, i, j) combines split and substring functions, with the input tokenized using character c and subsequently substr(i, j) is applied on the tth token.

One may question the significance of the splitSubstr unit when split and substr units already exist individually. The reason for these compound units is that transformations are formed as a concatenation of units, and stacking of the units is only possible by introducing new units (e.g., splitSubstr).

For a given set of source and target pairs (s_i, t_i), the coverage of a transformation is defined as the fraction of pairs in which applying the transformation on s_i generates t_i. When a transformation can produce more than one output, the coverage is defined as the fraction of pairs in which applying the transformation on s_i leads to a set that includes t_i. When transformations are employed to enable a join between two tables that are not otherwise equi-joinable, each transformation is applied to a row in the source, producing one or more candidates that are then looked up in the target.

4 Methodology

In this section, we introduce our generalization strategies and discuss how they are applied to a base set of transformations. We also analyze their impact on the coverage and the overall quality of the transformations.

4.1 Length Generalization Using Relative Indices

The invariance of a table transformation to input length is a desirable property, as it is expected to mitigate overfitting, especially if the input length varies between the extracted examples given as context and those used for testing the transformations. Many parameters in a transformation make references to input using some indices, and those indices can be absolute or relative. Relative indices need an anchor point, which may be set to the beginning, the end or any other input position. All transformations introduced in Auto-join and CST use absolute indices. Those indices can be replaced with relative indices without changing their semantics if we set the anchor point the beginning of the input. However, one can change the anchor point of a relative index to a different location, allowing the expression of transformations that will not be possible otherwise.

As possible anchor points, let s and e mark the beginning and the end of input respectively. Relative indices in a transformation can be stated with respect to either s or e, allowing two-way parsing of the input. For example, consider the first and second rows from the Course Catalog and Linkedin tables in our running example from Sect. 1, as depicted in Fig. 1. The transformation *<substr(0, 1), splitSubstr(‘ ’, 1, 0, 6), literal(‘@ualberta.ca’)>*, generated under CST, maps the first row but not the second because of the presence of a middle name. However, if we use relative indices, the transformation *<substr(s, s+1), splitSubstr(‘ ’, e, s, s+7), literal(‘@ualberta.ca’)>*, generated from the second row, will cover both rows.

4.2 Recurrence Generalization

Systematically handling repeating patterns is a generalization strategy that can reduce the complexity of integrating diverse data sources, improve the simplicity of mapping rules and potentially decrease the likelihood of errors in the integration process. In the context of table transformations, a unit that is applicable to an element in an input sequence may be applicable to all elements in the sequence. For example, consider a sequence of space-separated terms in a source column and a transformation that takes the first letter of each term and concatenates them together to produce the target. If the input length, in terms of the number of terms, is fixed at k for all rows, one can construct a sequence of k splitSubstr functions, each taking a term in the input sequence and outputting the first character of the term. However, this approach will not work if the input length varies from one row to the next. We introduce repeating patterns, in the form of unit repetition and unit removal, to address this problem.

Unit Repetition. Introducing repetition to transformation rules is not as straightforward as one may wish. This is because our transformation units use indices to indicate parts of input they operate on, and the behavior of those indices must be defined under repetition. For example, consider *split('', s)* which breaks the input based on the space character and takes the first element. Repeating this operation without changing the index will produce the same output and is less desirable. Since we want to break the input into smaller segments and operate on each segment through repetition, we will focus on generalizing the *split* and *splitSubstr* functions[3].

Definition 1. *We define split and splitSubstr under repetition as*

$$split(c,i)^r = \{split(c,i+j)|j \in [0,r-1]\},$$

$$splitSubstr(c,i,p,q)^r = \{splitSubstr(c,i+j,p,q)|j \in [0,r-1]\},$$

where r is the repetition factor, c is separator character(s) breaking the input into segments, and i, p, and q are indices, which can be both relative and absolute indices, and $i+j$ is constrained by the input length.

The concept of repetition can be applied to either a single unit or to a sequence of consecutive units within a transformation. As an example, consider the first and second rows from the Course Catalog and Publications tables in Fig. 1. The transformation *<splitSubstr('', s, s, s+1), literal ('. '), split('', e)>* maps the first row of the source table to the first row in the target table, but it cannot map the second row because of the presence of a middle name. However, the first two units can be applied multiple times under repetition. In particular, with at most two repetition of those two units, both rows are covered.

Unit Removal. As a special case of repetition, both being or not being a unit in the rule can generate valid outputs.

Definition 2. *We define split, substr, and splitSubstr under removal as*

$$split(c,i)^? = split(c,i)|\varnothing,$$

[3] *splitSubstr* becomes *substr* when the splitting character does not exist in the input, hence *substr* is not considered.

$$substr(p,q)^? = substr(p,q)|\varnothing,$$
$$splitSubstr(c,i,p,q)^? = splitSubstr(c,i,p,q)|\varnothing,$$

where ? represents the removal concept, and the separator character and indices remain as previously defined.

The transformation T_G applied to the second and third rows from the Course Catalog and Linkedin tables in Fig. 1 illustrates this concept, where transforming the third row becomes possible by removing a subpart of a previously generated transformation.

4.3 Source-Target Direction Generalization

When dealing with numerous tables from different sources, determining which table should serve as the source and which as the target in the mapping process can be challenging. Ideally, we prefer a more content-rich table to act as the source, providing ample evidence to construct the target. However, identifying such content-rich tables may not be straightforward. As an example, consider Course Catalog and Linkedin tables in Fig. 1. One heuristic for detecting the source involves selecting the more informative column based on length [18]. This implies choosing the column with longer text as the source and treating the column with shorter text as the target. While this strategy may work for some table pairs, it is evident that it would fail in selecting the source here. In this case, the email column is longer due to the literal at the end, causing it to be chosen as the source, whereas the name column is more descriptive.

One potential generalization is to avoid fixing the source and target and allow mapping in both directions. One can dynamically choose the direction that yields more general transformations. In our case, we take a small subset of each table as a sample and generate transformations based on this sample for both directions. The direction that can map rows with fewer transformations is deemed as the source, as those transformations are expected to exhibit better generalization.

4.4 Simplicity Generalization on Unseen Data

In many real-world scenarios, table rows undergo updates, and new rows are regularly added over time. In such dynamic environments, there is a need to continuously incorporate new data in the join process. However, repeatedly generating transformations from scratch each time new data arrives can be a time-consuming and resource-intensive task. A general transformation is expected to not only cover the existing rows, but also anticipate unseen rows that may arrive in the future. One approach to gauge the generality of a transformation is to estimate its coverage based on data observed thus far. A transformation that covers more rows is deemed more general. However, without comprehensive knowledge about the distribution of future data, it is generally challenging to say with certainty whether one transformation will offer greater coverage than another. A significant issue impacting the generality of transformations is the existence of accidental patterns in data. While such patterns may be present in a small sample, they are less likely to generalize to a larger sample or unseen data. Consequently, transformations constructed using a small sample may exhibit some of these accidental patterns. To mitigate such occurrences, we introduce simplicity as an additional measure of generality. In this context, simpler transformations are favored over more complex ones when all other factors, such as coverage on seen data, remain the same. In our case, a simplicity score may be defined in terms of either the number of units or the number of parameters in a transformation, where a transformation with fewer units or parameters is considered simpler. In our experiments, transformations are primarily chosen based on their coverage. In cases where there is a tie in coverage, a simpler transformation is favored over a more complex one. For example, consider two generated transformations with identical coverage on input data, where all their units are the same except for one. Suppose one transformation has a *substr(i, j)* unit, while the other includes *splitSubstr(c, 0, i, j)*. When character c is absent from the input data, both transformations have the same coverage. However, when considering transformation simplicity, we favor the former because if unseen data has a matching character c by chance, only *substr(i, j)* may perform the correct mapping.

5 Experiments

Our evaluation is conducted on two real-world datasets gathered from diverse sources with varying levels of noise and inconsistency, providing opportunities to evaluate various aspects of our generalization techniques. ***(1) Web Dataset*** [30] is generated by sampling table-related queries from the Bing search engine and retrieving sets of tables from Google Tables corresponding to those queries. The tables in this dataset are manually selected as those representing the same entities with different formatting that are joinable via textual transformations. This dataset consists of 31 pairs of web tables, covering 17 diverse topics. On average, each table contains approximately 92 rows, and the average length of a join entry is 31 characters. It is a challenging benchmark due to inconsistencies in the data and the presence of different formatting and patterns across the rows of each key column. ***(2) Spreadsheet Dataset*** [1] is taken from the public transformation benchmarks in FlashFill [8,9] and BlinkFill [22]. This benchmark is formed by the data cleaning challenges reported by the spreadsheet users on Microsoft Excel forums. The dataset contains 108 table pairs, each with 34 rows and 19 characters per join entry on average.

In our experiments, we set the maximum number of units and repetition degree to 3 and 2 respectively. We assume a manually annotated set of joinable row pairs is provided, as finding the matching pairs is beyond the scope of this study. Moreover, coverage values are averaged across all tables in the benchmark, and in cases where randomness affects the results, experiments are repeated at least 5 times, and the average is reported. Our baseline for comparison is CST [18], which, to the best of our knowledge, represents the state-of-the-art in finding explainable textual transformation.

Table 1. Best transformation coverage

Dataset	Baseline	Bidir.	Rel.	+Rem.	+ Rep.
Web	0.577	0.606 (2.90%)	0.676 (9.90%)	0.679 (10.21%)	0.679 (10.21%)
Spr.	0.667	0.667 (0.00%)	0.708 (4.09%)	0.725 (5.79%)	0.725 (5.79%)

(a) On all tables

Dataset	Baseline	Bidir.	Rel.	+Rem.	+ Rep.
Web	0.514	0.548 (3.33%)	0.628 (11.36%)	0.631 (11.72%)	0.631 (11.72%)
Spr.	0.308	0.308 (0.00%)	0.393 (8.49%)	0.429 (12.02%)	0.429 (12.02%)

(b) On tables without full coverage trans.

Table 2. The number of required transformations

Dataset	Baseline	Bidir.	Rel.	+Rem.	+ Rep.
Web	431	395 (8.35%)	396 (8.12%)	394 (8.58%)	394 (8.58%)
Spr.	970	970 (0.00%)	890 (8.25%)	876 (9.69%)	876 (9.69%)

Table 3. The impact of removal and repetition concepts

Dataset	+Rem.		+Rep.	
	Affected Trans.	Cov. Imp.	Affected Trans.	Cov. Imp.
Web	20.00%	5.90%	23.08%	6.48%
Spr.	8.08%	1.56%	8.08%	1.56%

5.1 Transformation Coverage

In this experiment, our aim is to generate transformations that cover all rows in the input. Since the framework is designed to generate a covering set, the coverage in our experiments is 100%, for both our approach and the baseline. To evaluate the impact of our generalization, we introduce two alternative metrics to measure the coverage of transformations: (1) *best transformation coverage*, which indicates the coverage achieved by the transformation that produces the highest coverage on the input rows; and (2) *the number of generated transformations*, because, when a smaller number of transformations is required to cover all input rows, it serves as an indicator that the transformations are more general. Table 1a summarizes the best transformation coverage when applying each generalization technique, as well as that of the baseline approach. Table 1b demonstrates the same evaluation, excluding tables that are covered by a single transformation. These excluded tables are considered easy, as a single transformation with non-general units may also achieve full coverage, leaving no room for generalization. Finally, Table 2 shows the total number of generated transformations required to achieve a coverage of 1.00 on all tables in the benchmark. The numbers inside parentheses indicate the improvement compared to the baseline. We analyze the effect of each generalization strategy individually:

Source-Target Direction Generalization: The "bidir." column in the aforementioned tables denotes the performance when applying our bidirectional generalization (Sect. 4.3). As demonstrated, our approach outperforms the baseline on the web dataset, which is a relatively more challenging dataset. This shows that the heuristic used in our baseline, which considers columns with lengthier rows as more context-rich, may not accurately label source and target columns. On the other hand, on the spreadsheet benchmark, our performance is on par with the baseline, since in this dataset, the source is always lengthier.

Length Generalization: The utilization of relative indices considerably enhances the performance of our approach compared to the baseline, as observed across all metrics on both datasets. The "Rel." column in Tables 1 and 2 quantifies this improvement. In each dataset, there are several tables covered with more transformations in the baseline due to the limitations of absolute indices in modeling length variance in the rows. Applying relative indices results in the generation of more general transformations, capable of covering more rows compared to their absolute counterparts. This can lead to up to a 10% improvement in the coverage of the best transformation and an 8% reduction in the number of required transformations, particularly on the challenging web dataset.

Recurrence Generalization: Columns "+Rem." and "+Rep." denote the performance when augmenting relative-indexed transformations with unit removal and unit removal and repetition, respectively. As shown, there is a slight improvement in both the best transformation coverage and the number of required transformations when employing these generalizations. However, the best transformation coverage and the number of required transformations may not give the full picture when in fact our generalizations considerably enhance the coverage of transformations. Therefore, instead of relying solely on the best transformation coverage, we measure the improvement for each transformation individually, summarized in Table 3. For this evaluation, we compare the coverage of each transformation when only generalized by relative indices with those that are also augmented by recurrence generalization. Transformations with 100% coverage and those covering only a single row (i.e., typically literals) are excluded in this evaluation, as they cannot be generalized further. As shown, on the web dataset, both unit removal (denoted by "+Rem.") and unit removal and repetition (denoted by "+Rep.") generalize 20% or more of the transformations, and the coverage is increased by about 6%, which underscores the effectiveness and importance of recurrence generalization methods.

5.2 Generalization to Unseen Data

The improved coverage on unseen data in our approach is primarily achieved through length and simplicity generalization techniques. In our first experiment, we aim to measure the extent to which simplicity generalization can impact the generated transformations on each dataset. Specifically, we conducted the experiment using two tie-breaking strategies when transformations with equal coverage are to be selected. In one strategy, a simple transformation is prioritized over a complex one, and in the other strategy, the selection is made randomly. Our metrics for simplicity in one setting is the total number of units and in another setting is the total number of parameters. Simplicity-based tie-breaking results in at least a 5% reduction in the total number of units and a noticeable 17% decrease in the parameters needed. Not only do the simpler transformations lead to a faster join on large dynamic environments and provide an easier-to-understand mapping, but, as confirmed by the next experiment, they also yield better coverage on unseen data.

In the next experiment, we randomly divide the rows of each table into two sets. One set represents the available data based on which transformations are built, and the other set represents unseen data that may be added

Table 4. The impact of generalization on transformation coverage when limited data is available

Dataset	Sample	Baseline	General Trans.	Gap
Web	10%	0.528	0.590	0.061
	20%	0.663	0.704	0.041
	40%	0.794	0.800	0.005
Spr.	10%	0.634	0.640	0.005
	20%	0.681	0.682	0.001
	40%	0.793	0.801	0.008

(a) Tie-breaking: random

Dataset	Sample	Baseline	General Trans.	Gap
Web	10%	0.537	0.645	0.108
	20%	0.670	0.727	0.057
	40%	0.799	0.816	0.017
Spr.	10%	0.639	0.675	0.035
	20%	0.684	0.710	0.026
	40%	0.797	0.810	0.012

(b) Tie-breaking: simplicity-based

in the future. We use 10%, 20%, and 40% of the rows as the available data. For tie-breaking among transformations with equal coverage, two strategies are evaluated: simplicity-based and random selection. Table 4a shows the performance using random selection, and Table 4b indicates the same experiment using simplicity-based selection, for both the baseline approach and ours.

Two important observations can be made: (1) Regardless of the tie-breaking strategy employed, our approach consistently outperforms the baseline when measuring coverage on unseen data. The performance gap between the baseline and our approach widens when a smaller set of data is available for transformation finding. This implies that, compared to the baseline, our general transformations demonstrate superior performance when generated from a smaller set of available data. Notably, the gap is more pronounced in the web dataset, reflecting the dataset's inherent complexities. (2) Employing simplicity-based tie-breaking enhances the coverage of both our approach and the baseline. Nonetheless, our approach benefits more from this generalization, and the improvement is more pronounced in the web dataset, particularly for smaller sample sizes.

5.3 End-To-End Join Performance

In the final set of our experiments, we evaluated the performance of our approach on an end-to-end join between two tables with unequal column values. To employ transformations for joinability, we set the repetition degree to one, which ensures a single output for each transformation applied on an input. In addition to the automated similarity-based example generation in CST (referred to as *Automated Example Generation* in the rest of this section), we also experimented with a small set of 5 manually prepared examples for each table, while the remaining rows were kept for testing, denoted as *Manual Example Generation.* Table 5 summarizes the unequal join performance in terms of Precision, Recall, and F1-Score (denoted with P, R, and F1, respectively) for our approach, CST, and another transformation-based joining approach, Auto-join[4]. Due to the limitations of the sampling framework and subset-based generation in Auto-join, it was only employed when examples were automatically generated and, thus, large enough to be divided into subsets.

[4] The experiments on Auto-join were conducted using the implementation by the authors of CST [18].

Table 5. Join performance of our approach and baselines

	Our Approach			CST			Auto-Join		
	P	R	F1	P	R	F1	P	R	F1
Web	0.953	0.754	0.777	0.879	0.726	0.713	0.985	0.415	0.466
Spr.	0.997	0.787	0.810	0.995	0.792	0.812	0.997	0.768	0.796

(a) Automated example generation

	Our Approach			CST		
	P	R	F1	P	R	F1
Web	0.993	0.612	0.672	0.992	0.571	0.631
Spr.	0.986	0.648	0.663	0.986	0.614	0.624

(b) Manual example generation

As demonstrated, regardless of the sampling method, our approach outperforms all baselines by a noticeable margin in the web dataset, yielding up to an 8% improvement in F1-Score. The diversity and complexity of the patterns in input rows in this dataset better highlight the ability of our approach to extract more general transformations and, consequently, achieve a better join performance while maintaining the explainability of the mapping space. When examples are automatically generated in the spreadsheet dataset, our performance is on par with the baseline. This is because the dataset is clean, the rows are consistently covered by a few rather easy textual patterns, and the example generation framework returns a sufficient number of correct examples. On the other hand, as shown in Table 5b, the performance gap exists for both datasets when examples are limited to a few human-generated ones. This gap is expected, considering the better generalization of our approach to unseen data[5].

6 Conclusion

We have studied the problem of joinablity for non-equi-joinable columns, with a focus on explainability and the generality of transformations. Our exploration has encompassed diverse generalization techniques, such as length invariance, pattern repetition, bi-directionality and simplicity. Our evaluation highlights that our method not only produces more effective transformations but also offers a higher degree of simplicity and explainability compared to competitors.

Acknowledgments. This research was supported by the Natural Sciences and Engineering Research Council and the PS752 commemorative scholarship program.

References

1. Alur, R., Fisman, D., Singh, R., Solar-Lezama, A.: SyGuS-comp 2016: results and analysis. arXiv:1611.07627 (2016)
2. Bonifati, A., Comignani, U., Coquery, E., Thion, R.: Interactive mapping specification with exemplar tuples. ACM TODS **44**(3) (2019)

[5] Our observations indicate that the best performance-recall trade-off is achieved when the minimum support for transformations is set to 0.05, and all experiments in this section are conducted with that threshold.

3. Cafarella, M.J., Halevy, A., Wang, D.Z., Wu, E., Zhang, Y.: Webtables: exploring the power of tables on the web. Proc. VLDB Endow. **1**(1) (2008)
4. Cate, B.T., Kolaitis, P.G., Qian, K., Tan, W.C.: Approximation algorithms for schema-mapping discovery from data examples. ACM TODS **42**(2) (2017)
5. Chaudhuri, S., Ganti, V., Kaushik, R.: A primitive operator for similarity joins in data cleaning. In: 22nd ICDE (2006)
6. Chen, X., Liang, C., Yu, A.W., Song, D., Zhou, D.: Compositional generalization via neural-symbolic stack machines. In: NeurIPS, vol. 33 (2020)
7. Conklin, H., Wang, B., Smith, K., Titov, I.: Meta-learning to compositionally generalize. arXiv:2106.04252 (2021)
8. Gulwani, S.: Automating string processing in spreadsheets using input-output examples. ACM Sigplan Not. **46**(1) (2011)
9. Gulwani, S., Harris, W.R., Singh, R.: Spreadsheet data manipulation using examples. Commun. ACM **55**(8) (2012)
10. He, J., et al.: Interactive and deterministic data cleaning. In: Proceedings of the 2016 International Conference on Management of Data (2016)
11. Heer, J., Hellerstein, J.M., Kandel, S.: Predictive interaction for data transformation. In: CIDR (2015)
12. Herzig, J., Berant, J.: Span-based semantic parsing for compositional generalization. arXiv:2009.06040 (2020)
13. Lake, B.M.: Compositional generalization through meta sequence-to-sequence learning. In: NeurIPS, vol. 32 (2019)
14. Li, Y., Li, J., Suhara, Y., Doan, A., Tan, W.C.: Deep entity matching with pre-trained language models. Proc. VLDB Endow. **14**(1) (2020)
15. Ling, X., Halevy, A.Y., Wu, F., Yu, C.: Synthesizing union tables from the web. In: Twenty-Third International Joint Conference on Artificial Intelligence (2013)
16. Liu, Q., et al.: Compositional generalization by learning analytical expressions. In: NeurIPS, vol. 33 (2020)
17. Nargesian, F., Zhu, E., Pu, K.Q., Miller, R.J.: Table union search on open data. Proc. VLDB Endow. **11**(7) (2018)
18. Nobari, A.D., Rafiei, D.: Efficiently transforming tables for joinability. In: 2022 IEEE 38th ICDE. IEEE (2022)
19. OpenRefine: Openrefine documentation: Cell editing (2010). https://openrefine.org/docs/manual/cellediting. Accessed 14 Jan 2024
20. Russin, J., Jo, J., O'Reilly, R.C., Bengio, Y.: Compositional generalization in a deep seq2seq model by separating syntax and semantics. arXiv:1904.09708 (2019)
21. Shi, K., Hong, J., Zaheer, M., Yin, P., Sutton, C.: Compositional generalization and decomposition in neural program synthesis. arXiv:2204.03758 (2022)
22. Singh, R.: BlinkFill: semi-supervised programming by example for syntactic string transformations. Proc. VLDB Endow. **9**(10) (2016)
23. Wang, B., Lapata, M., Titov, I.: Meta-learning for domain generalization in semantic parsing. arXiv:2010.11988 (2020)
24. Wang, J., Li, G., Fe, J.: Fast-join: an efficient method for fuzzy token matching based string similarity join. In: 2011 IEEE 27th ICDE (2011)
25. Wang, J., Li, G., Feng, J.: Extending string similarity join to tolerant fuzzy token matching. ACM Trans. Database Syst. **39**(1) (2014)
26. Zhang, L., Zhang, S., Balog, K.: Table2vec: neural word and entity embeddings for table population and retrieval. In: Proceedings of the 42nd International ACM SIGIR Conference on Research and Development in Information Retrieval. ACM, New York (2019)

27. Zhang, S., Balog, K.: Semantic table retrieval using keyword and table queries. ACM Trans. Web **15**(3) (2021)
28. Zhao, C., He, Y.: Auto-EM: end-to-end fuzzy entity-matching using pre-trained deep models and transfer learning. In: The World Wide Web Conference, WWW 2019. ACM (2019)
29. Zhu, E., Deng, D., Nargesian, F., Miller, R.J.: Josie: overlap set similarity search for finding joinable tables in data lakes. In: Proceedings of the 2019 International Conference on Management of Data, SIGMOD 2019. ACM (2019)
30. Zhu, E., He, Y., Chaudhuri, S.: Auto-join: joining tables by leveraging transformations. Proc. VLDB Endow. **10**(10) (2017)
31. Zhu, E., Nargesian, F., Pu, K.Q., Miller, R.J.: LSH ensemble: Internet-scale domain search. Proc. VLDB Endow. **9**(12) (2016)

Density Based Learned Spatial Index for Clustered Data

Xiaofei Zhao(✉) and Kam-Yiu Lam

City University of Hong Kong, Kowloon Tong, Hong Kong
xf.zhao@my.cityu.edu.hk, cskylam@cityu.edu.hk

Abstract. Retrieving spatial points, such as GPS records, that meet specific location criteria from a spatial database is a crucial operation in location-based services. Recent research suggests that learned indexes can surpass traditional ones in both query performance and space efficiency by leveraging data distribution information to construct a compact model, whereas traditional indexes make minimal assumptions about data distribution. In real-world spatial databases, spatial data often clusters, indicating a non-uniform distribution but concentration in specific regions or along road networks. Adaptivity of the index to this data pattern is crucial in such scenarios. In this paper, we discuss building efficient learned indexes by capitalizing on the clustering property of the dataset. Specifically, we propose a Density-based Grid Learning Spatial Index (DGLSI) that partitions spatial data based on data point density and employs learned models, including spatial interpolation functions and a recursive model index, to predict the grid cell IDs of query points. We evaluate DGLSI's performance on real-world GPS datasets and demonstrate that the proposed methods outperform analogous grid-based indexes across various query workloads, including nearest point queries and range queries while maintaining high space efficiency.

Keywords: learned index · spatial index · clustered data · density-based partitioning

1 Introduction

Spatial data indexing, such as efficiently locating spatial points in large spatial databases, is a critical component of location-based services (LBS) and geographic information systems (GIS). Traditional spatial indexing techniques like R-trees [4] and space-filling curves (SFCs) [1] have well-known trade-offs between flexibility in accommodating data distributions and efficiency in locating exact spatial points. Recent works [7,9–11] have explored using learned models for spatial indexing to take advantage of the underlying data distribution.

While the initial learned indexes [5] are designed for one-dimensional data, intuitively, multidimensional data like spatial points can be transformed into a one-dimensional space using an SFC, such as a Z-curve, and then indexed by

J. Tekli et al. (Eds.): ADBIS 2024, LNCS 14918, pp. 138–151, 2024.
https://doi.org/10.1007/978-3-031-70626-4_10

a learned model. Several works like ZM index [9], SPRIG [10], and LPBF [12] utilize this idea. However, we have not seen sufficient discussions on how the efficiency of these methods is affected by the data distribution and the chosen model parameters. For example, the ZM index uses the Z-curve to encode the spatial points, allowing the dataset to be sorted by the Z-address. Given a spatial point, the Z-address can be calculated in constant time. Then, a learned model is used to predict the storage position of the point in the dataset. When applying an SFC for spatial indexing, it is necessary to carefully handle the locality properties [3] of the curve. In some subspaces of the service area, two spatial points p_1 and p_2 that are close to each other in the 2-D space may be far away from each other in the transformed 1-D space. This may lead to significant degradation in the performance of a range query.

This study proposes a new learned spatial index called the Density Based Learned Spatial Index (DGLSI) that aims to combine the strengths of traditional approaches and learned techniques, e.g., the adaptivity to the data distribution like R-tree, and the efficiency in query performance and index storage from learned indexes. DGLSI is designed specifically for handling unbalanced, clustered spatial data distributions that commonly occur in real-world applications, such as points of interest (PoIs) or GPS trajectories. By utilizing density-based partitioning and learned models, DGLSI can leverage the advantages of traditional indexing approaches while addressing their limitations when dealing with non-uniform data. The key ideas behind DGLSI are: (1) Partitioning spatial data into variably-sized grid cells based on data density rather than using a uniform grid. (2) Sorting grid cells using a ZigZag curve to assign each cell a unique, sortable ID. (3) Employing one-dimensional learned models to predict the row and column IDs of the grid cell containing the spatial point requested in a spatial query. The remainder of this paper is organized as follows. Section 2 covers the background concepts and existing approaches. Section 3 details the proposed DGLSI index structure, data layout, index building process, query processing algorithms, and spatial interpolation models. Section 4 presents an evaluation that analyzes the storage overhead, query performance, and other trade-offs of DGLSI compared to prior methods.

2 Background

2.1 Learned Index

Traditional indexes do not assume specific data distributions. Leveraging distribution information can reduce storage overhead without significantly increasing search time. For example, in a simple key-value data storage with keys as continuous integers from 1 to N and fixed-size data entries, traditional index structures like B-Trees may require $\mathcal{O}(N)$ additional nodes to index the entire dataset. On the other hand, a learned index can "predict" the position of any record with an $\mathcal{O}(1)$ space overhead. In this case, a simple linear function can serve as the learned index model: $pos = M(k) = wk + b$, where k is the look-up key, and w, b are model parameters to be trained on the dataset. Existing studies [5] have shown the efficiency of the learned index in saving storage overhead for even more complex data distributions without compromising query performance in many cases, compared to traditional index structures like B-Tree and hash index.

In this paper, we adhere to the definition of an index, which requires that it exhibits no false negatives. In other words, if a data record exists in the data storage, the index must accurately locate and retrieve that data record upon a query request. However, learned models may make mistakes. In the context of sorted datasets, inaccuracies in learned models are typically manageable, as most learned index architectures incorporate a local search (a.k.a. "last mile search") after the initial prediction, thereby ensuring the correctness of the result.

2.2 Spatial Point Indexes

The literature has demonstrated the efficiency of learned indexes, primarily under the assumption that the keys' order is well-defined, and the dataset can be stored in a sorted linear data structure. This sorted nature of the dataset ensures correctness and bounds the cost of the last-mile search. When dealing with spatial data, or more generally, multidimensional data, the definition of "the order of data entries" can be ambiguous. People design spatial index structures based on the expected frequently occurring spatial queries, such as nearest point queries and spatial range queries. These spatial index structures tend to store proximate points closely in the database.

For instance, one of the widely used spatial index in popular spatial databases, the R-tree, is based on the bounding box of spatial objects and operates on the principle of branch and bound (BB) [6]. To construct an R-tree index, the service area is hierarchically divided into smaller areas, usually rectangles, organized in a tree structure. On the other hand, when using SFC for spatial data indexing, the service area is typically divided into a hierarchy of cell grids by recursively splitting the area into four equal parts. The process of splitting a cell into four smaller cells at the next level can be interpreted as appending two bits to the cell ID. Utilizing an SFC-based index theoretically allows for achieving very high resolution, implying that leaf-level cells can be extremely small and contain only a few data entries.

A limitation of SFC-based spatial indexes is their inability to adaptively partition cells based on data distribution. Consequently, the number of data entries in a cell can vary significantly, posing challenges when determining the appropriate level of the cell grid. If leaf-level cells are too large, query performance may degrade due to the large number of data entries in each cell. Conversely, excessively small cells can lead to examining numerous cells (with discontinuous cell IDs) during a spatial range query. In summary, the R-tree-based index provides flexibility in determining the spatial range of each node based on data distribution, whereas the SFC-based index excels in efficiently locating exact spatial points owing to its regular cell grid structure.

2.3 Clustered Data Distribution

In real-world LBS and spatial databases, spatial data points are typically non-uniformly distributed. Figure 1a displays a set of randomly generated, uniformly distributed spatial points, while Fig. 1b illustrates a real-world example of the San Francisco road network nodes obtained from OpenStreetMap. In many scenarios involving PoIs or vehicle GPS trajectories, the points tend to cluster significantly in certain areas, such as along roads or in city centers. In such cases, the precision of a uniform grid-based index may degrade.

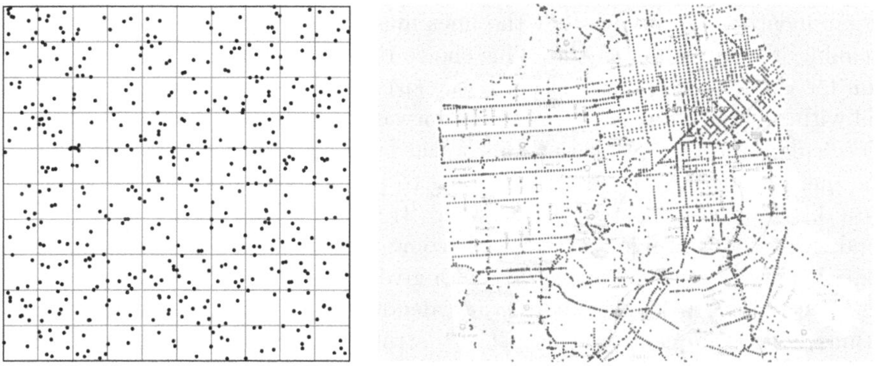

(a) Uniform (b) Clustered

Fig. 1. Different data distributions

3 The Proposed Index Structure: DGLSI

In this section, we delve into the design and efficiency of a spatial interpolation function-based learned index for clustered spatial data. Our goal is to create a learned spatial index that efficiently locates spatial points similar to the SFC-based index, while also being adaptive in partitioning the data based on the data distribution.

- *The dataset.* We assume the given dataset $\mathcal{D}$ comprises 2-D spatial points $p_1, p_2, \ldots, p_n$, where $p_i = (x_i, y_i)$. In real-world applications, the dataset may consist of PoIs or GPS trajectories.
- *The service area.* The service area is a rectangle with a Cartesian coordinate system. Without loss of generality, we assume the bottom-left corner of the service area is $(-180, -90)$. When dealing with real-world spatial data, selecting an appropriate spatial reference system (SRS) to represent the spatial points is typically necessary, although it is beyond the scope of this paper.

3.1 Data Layout

The fundamental unit of the spatial data storage system is represented as a string type, termed a `record_t`. Each `record_t` encodes the coordinates (x, y) of a single spatial point along with other data attributes. Additionally, we assume that the size of each record is pre-defined and fixed. This assumption is reasonable, as when constructing a spatial data storage system for, for instance, GPS trajectory data, the format of each received GPS point record is typically pre-defined.

The service area is divided into a cell grid, as depicted in Fig. 1a. It's important to note that the spacing of the vertical and horizontal lines in the grid is not necessarily consistent (a.k.a. no need to be a Cartesian grid). The only requirement for the grid is that the lines must be parallel to the x and y axes, meaning it is a rectilinear grid. This choice is derived from the previous discussion for the following reasons: (1) If the size of each cell is fixed, indexing the cell with a space-filling curve is straightforward, and the time complexity of cell ID calculation is $\mathcal{O}(1)$. However, this layout lacks adaptivity to data distribution and query workloads, potentially leading to performance degradation when the data distribution is clustered or skewed. (2) Conversely, if the grid is irregular, conducting the last mile search may become infeasible because determining the spatial range and neighbouring cells of a given cell becomes challenging.

Each cell is considered a *data group*, denoted as $G_{i,j}$ in our subsequent algorithms, and is implemented as a C++ structure called `group_t`. A `group_t` comprises the following attributes: (1) the spatial range of the group, defined by the lower-left and upper-right corners of the cell; (2) the number of data entries in the group; (3) the data entries (`record_t`) stored in the group, which are stored in a continuous array and sorted by their first-dimensional coordinates. For persistent storage, multiple SPSVERBc3s are then serialized into a file, sorted by the cell ID. There is a `manifest` file containing the metadata of the storage system, including a table that maps the cell ID to the byte offset of the corresponding `group_t` in a file. Upon initialization of DGLSI, the `manifest` file is loaded into memory, and the mapping table is constructed in memory.

3.2 Query Types

The proposed method aims to support the following two types of spatial queries:

1. *Nearest neighbor query* $Q(p_c, r)$: Given a query point p_c and an optional parameter r, this query seeks the spatial point p nearest to p_c with $d(p, p_c) \leq r$, where $d()$ is the distance function. The query result should return the spatial point p. This query type can be readily extended to the k-nearest neighbor query.
2. *Range query* $Q(p_1, p_2)$: Given two points p_1 and p_2 defining a rectangle in the 2-D space, this query retrieves all spatial points falling into the rectangle. The query result should be a set of spatial points.

3.3 Index Building

Preparation. Before the index-building process, scanning the dataset is necessary to collect information about the data distribution. This preparation process generates two sorted arrays: X containing the first-dimensional coordinates and Y containing the second-dimensional coordinates of the data points.

Density Based Partitioning. Given the dataset $\mathcal{D}$, the initial step of building the index structure involves partitioning the service area into a grid. This partitioning process aims to achieve a more balanced distribution of data entries in each cell. To accomplish this, we establish a threshold N_c for the number of data records in each column and row of the grid. N_c serves to limit the number of data records in each grid column and row, ensuring efficient last-mile searches. For instance, if we aim to cap the average number of data records in each cell at 10, then N_c can be estimated as $\sqrt{\frac{\text{total number of records}}{10}} \times 10$.

The partitioning algorithm takes the sorted arrays X and Y as input and produces two sorted arrays $B_x = [b_1^x, b_2^x, \ldots, b_m^x]$ and $B_y = [b_1^y, b_2^y, \ldots, b_n^y]$ that define the boundaries of the cells in the grid (as depicted in Fig. 2). The partitioning algorithm is given as a recursive function called `Partition()`. The function takes the reference to the sorted array X and the threshold N_c as input; additionally, it takes the left and right indices (for X) of the interval $[l, r)$ that the function is currently working on. The function partitions the interval $[l, r)$ into two equal sub-intervals. By "equal", we mean that the number of data points in the left sub-interval is equal to that in the right sub-interval. Given the middle index m, we find the value $X[m]$ and insert it into the resulting boundary array B_x. By partitioning the interval recursively until the number of data points in each sub-interval is less than N_c, we get the final boundaries B_x. Algorithm 1 gives the example for getting the vertical boundaries B_x.

As depicted in Fig. 2, utilizing the boundaries B_x and B_y, we "virtually" partition the service area into a grid. The density of grid lines adapts to the data distribution. Finally, we employ a straightforward one-dimensional curve (not necessarily a space-filling curve) to allocate a cell ID to each grid cell. Following the approach in [10], we utilize the Zig-zag shaped curve (not Z-curve, the SFC) for indexing the grid, ensuring that each cell receives a unique, monotonically increasing cell ID along each dimension.

Storing Data Groups. After virtually partitioning the service area by computing the boundary arrays B_x and B_y, we must physically organize the original spatial data records into the designated data groups. Essentially, we need to scan the dataset again, and for each data record p_i, locate the corresponding `group_t` $G_{x,y}$. Since the cell sizes are not necessarily equal, we cannot directly compute the cell ID of p_i from its coordinates. Instead, we must search the boundary arrays B_x and B_y to find the cell ID of p_i. This process essentially becomes a one-dimensional search problem, where we can construct two learned indices on

Algorithm 1. GetBoundary(X, N_c)

Require: X, the sorted array of first-dimension coordinates of all data points; $X.max$, the maximum value in array X; N_c, the threshold of the number of data records in each column of the grid.

Ensure: B_x, a sorted array that defines the vertical boundaries in the grid.

$B_x \leftarrow [0, X.max]$
function PARTITION(X, N_c, l, r)
  if $r - l \leq N_c$ **then**
    return
  end if
  $m \leftarrow \lfloor \frac{l+r}{2} \rfloor$
  $B_x.insert\_sorted(X[m])$
  PARTITION(X, N_c, l, m)
  PARTITION(X, N_c, m, r)
end function
PARTITION($X, N_c, 0, X.size$)
return B_x

the boundary arrays B_x and B_y to predict the cell ID of a spatial point. We will delve into the one-dimensional learned index in the subsequent section.

Here, we assume that the cell ID of p_i is found and denoted as C_i. DGLSI checks the memory and the `manifest` file to determine if the group $G_{x,y}$ is already created:

- If the group is not yet created, a new group is generated with the spatial range derived from the boundaries B_x and B_y with the group's cell ID C_i. Subsequently, the data record p_i is inserted into the group. The group remains in the memory cache until it reaches capacity or has not been accessed for an extended period. At that point, the group is serialized into the corresponding file.
- If the group already exists in the system, it is loaded into the memory cache (if not already present), and the data record p_i is inserted into the group.

Algorithm 2 presents the pseudocode for the final step of the index-building process.

3.4 One-Dimensional Learned Index

The concept of the one-dimensional learned index is quite straightforward: using the boundary array B_x, we construct a two-stage Recursive Model Index (RMI) [5], denoted as $M_x()$, similar to the approach employed in [5]. Given a query point $p_i = (x_i, y_i)$, the objective is to locate the interval $[b_l^x, b_r^x)$ in B_x that contains x_i. Here, the input of the model $M_x()$ is x_i, and the prediction is the index of an element in the array B_x, denoted as idx. Subsequently, DGLSI inspects the value of $B_x[idx]$ and its neighboring values, such as $B_x[idx - 1]$ and $B_x[idx + 1]$, to ascertain the column ID of p_i. Although the RMI model provides

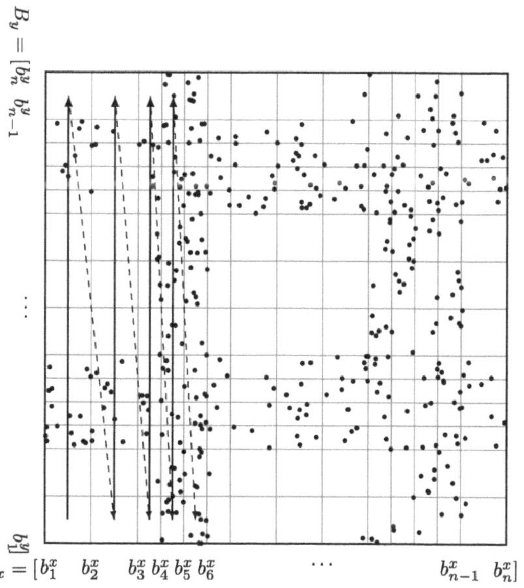

Fig. 2. Clustered data distribution

Algorithm 2. Store($\mathcal{D}, B_x, B_y$)

Require: $\mathcal{D}$, the dataset of spatial points; B_x, B_y, the boundary arrays that define the grid; $\mathcal{G}$, the set of data groups, a.k.a. cells, that has already been created; M_c, the memory cache that stores several data groups.

$M_c \leftarrow \{\}$
for $p \in \mathcal{D}$ **do**
  $C_i \leftarrow$ FindCellID(p, B_x, B_y)
  if $G_{C_i} \notin \mathcal{G}$ **then**
    $G_{C_i} \leftarrow$ CreateGroup(C_i)
    InsertIntoCache(G_{C_i}, M_c)
  end if
  if $G_{C_i} \notin M_c$ **then**
    $G_{C_i} \leftarrow$ LoadGroup(C_i)
    InsertIntoCache(G_{C_i}, M_c)
  end if
  $G_{C_i}.insert(p)$
end for
function InsertIntoCache(G, M_c)
  if $M_c.size > cache\_size$ **then**
    $G_{lru} \leftarrow$ FindLRU(M_c)
    G_{lru}.Serialize(())
    $M_c.remove(G_{lru})$
  end if
  $M_c.insert(G)$
end function

a starting point for the search, the search for the exact position of x_i continues until the interval $[b_l^x, b_r^x)$, where $b_l^x \leq x_i < b_r^x$, is identified.

The process for determining the y-dimensional cell ID in B_y mirrors the x-dimensional case. Once the column and row IDs are identified, the cell ID of p_i is computed using a formula representing the ZigZag curve:

$$C_i = col\_id \times num\_rows + row\_id. \tag{1}$$

where col_id and row_id are the identified column and row IDs, respectively, and num_rows is the number of rows in the grid.

3.5 Query Processing

Nearest Neighbor Query. Given the query $Q(p_c, r)$, the process of finding the nearest neighbor of p_c unfolds as follows:

1. Determine the cell ID of p_c using the one-dimensional learned indices.
2. Load the data group $G_{x,y}$ containing p_c into memory cache.
3. Execute a linear search in $G_{x,y}$ to identify the nearest neighbor of p_c.
4. Note that the nearest neighbour of p_c may not necessarily reside in the same cell as p_c. In such cases: (1) Assume there exists a data point p_i in the same cell as p_c and $d(p_i, p_c) \leq r$, which is the nearest to p_c within the cell. Drawing a circle centred at p_c with radius $d(p_i, p_c)$, if the circle intersects with the cell boundary, consideration of neighbouring cells becomes necessary since there might be a closer point than p_i in the neighbouring cells. (2) If only one data point p_c exists in the cell, all eight surrounding cells must be loaded and a search conducted within them. If at least one neighbouring point is found, the scenario reverts to the previous case.

Algorithms 3 and 4 present the pseudocode for processing the nearest neighbour query. Algorithm 3 ultimately identifies the nearest neighbour of p_c within the so-called "XGroup" (signifying the extended group encompassing an original group and its eight surrounding groups). Subsequently, Algorithm 4 extends the search area further if the distance between p_c and the nearest neighbour discovered exceeds the distance between p_c and the boundaries of the XGroup. Here, $XG.s$, $XG.n$, $XG.w$, and $XG.e$ denote the south, north, west, and east boundaries of the XGroup, respectively.

Range Query. Given the query $Q(p_1, p_2)$, the process of identifying all the spatial points that lie within the rectangle defined by p_1 and p_2 unfolds as follows:

1. Determine the cell IDs of p_1 and p_2 using the one-dimensional learned indices.
2. Compute the range of cell IDs falling within the rectangle defined by p_1 and p_2.
3. Load the corresponding data groups into the memory cache and conduct a linear search in each group.

Algorithm 3. NearestNeighbor(p_c, r), Part I

Require: p_c, the query point; r, the radius of the query.

$C \leftarrow$ FindCellID(p_c, B_x, B_y)
$G_{x,y} \leftarrow$ LoadGroup(C)
$XG \leftarrow G_{x,y}$
$d = r$
$p \leftarrow$ LinearSearch($G_{x,y}, p_c, r$)
$d = \min(d, d(p, p_c))$
while p = NULL **do**
    $N \leftarrow$ FindNeighbors(XG)
    for $C' \in N$ **do**
        $G_{x',y'} \leftarrow$ LoadGroup(C')
        $p' \leftarrow$ LinearSearch($G_{x',y'}, p_c, r$)
        if $p' \neq$ NULL **then**
            $d' = d(p', p_c)$
            if $d' < d$ **then**
                $d = d'$
                $p = p'$
            end if
        end if
    end for
    $XG \leftarrow XG \cup \{\text{neighbors}\}$
end while

4. (1) For groups partially intersected by the query range, inspect each data record in the group to ascertain if it falls within the query range. (2) For groups entirely contained within the query range, add all the data records to the result set.

4 Evaluation

4.1 Dataset and Experimental Setup

We employ the Geolife dataset version 1.3, curated and published by researchers from Microsoft Research, which encompasses GPS trajectories of 182 users over a span of more than five years (from 2007 to 2012). The majority of data points are situated in the city centre of Beijing, China. With 17,621 trajectories containing a total of 24, 876, 978 spatial points, the dataset reflects a non-uniform distribution owing to the clustering along the road network. During index construction, we set the anticipated average number of data records in each cell to 10.

The experiments are conducted on a Desktop PC equipped with an Intel Core i7-9700K CPU, 64 GB RAM, and a Samsung 980 Pro 1 TB NVMe SSD. We measure the storage overhead and query performance of the proposed index structure, contrasting it with the analogous grid-based index SPRIG [10]. DGLSI is implemented in C++, alongside a simplified version of SPRIG, featuring identical `record_t` structure and `manifest` file (housing the mapping table from cell

Algorithm 4. NearestNeighbor(p_c, r), Part II

Require: p_c, the query point; r, the radius of the query; p, the "currently" found nearest neighbor of p_c; XG, the XGroup that contains p_c and p. d, the distance between p and p_c.

while $d(p_c, p) >$ any of $d(p_c, XG.s), d(p_c, XG.n), d(p_c, XG.w), d(p_c, XG.e)$ **do**
    $N \leftarrow$ FINDNEIGHBORS(XG)
    for $C' \in N$ **do**
        $G_{x',y'} \leftarrow$ LOADGROUP(C')
        $p' \leftarrow$ LINEARSEARCH($G_{x',y'}, p_c, r$)
        if $p' \neq$ NULL **then**
            $d' = d(p', p_c)$
            if $d' < d$ **then**
                $d = d'$
                $p = p'$
            end if
        end if
    end for
    $XG \leftarrow XG \cup \{\text{neighbors}\}$
end while

ID to the byte offset of the corresponding group in the data file). The primary distinctions between DGLSI and SPRIG lie in the employed learned index structure (per-dimensional vs. spatial interpolation function) and data partitioning strategy. Before delving into the evaluation outcomes, we provide a brief overview of SPRIG.

4.2 Baseline Model: Simplified SPRIG

SPRIG employs the spatial interpolation function as the learned model to predict the cell ID of a spatial point. Spatial interpolation is a widely used method in geographic information systems (GIS) and image processing. It estimates the value of a variable at any location in a service area based on the measured values at observed locations [2,8]. Given a set of 2-D observed points $p_1, p_2, \ldots, p_n$, where $p_i = (x_i, y_i)$, and the corresponding values $v_1, v_2, \ldots, v_n$, the spatial interpolation function $f(x, y)$ estimates the value at a location (x, y) based on the observed values.

Figure 3 illustrates the bilinear interpolation function $f(x, y)$ used in SPRIG. There are the four coefficients to be trained on the observed data points. The four coefficients can be calculated by solving a linear system of equations given the values at the four corners $f(x_1, y_1), \ldots f(x_2, y_2)$. Given the established grid and the "Zig-Zag" cell ID numbering scheme illustrated by Fig. 2 and in Subsect. 3.4, if we use the cell IDs as the "observed value" and the cell boundaries as the function input, the bilinear interpolation function can be estimated using the sum of squared errors (SSE) as the loss function.

Using the bilinear interpolation function, the query process is quite straightforward: given a query point $p_i = (x_i, y_i)$, we estimate the cell ID of p_i

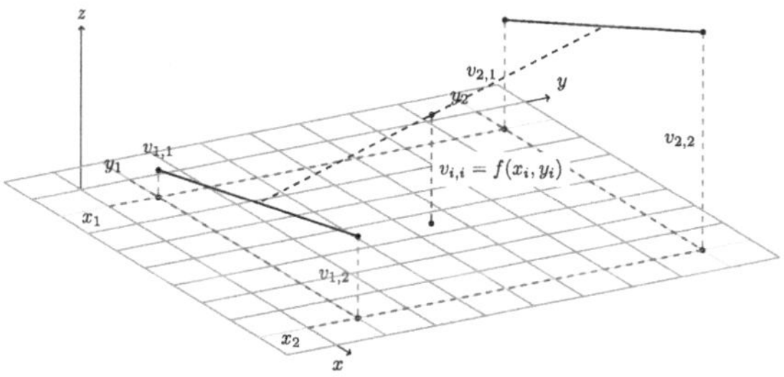

Fig. 3. Bilinear interpolation function

using the interpolation function $C_{pred} = f(x_i, y_i)$. Let $C_{pred,x}$ and $C_{pred,y}$ be the predicted column and row IDs, respectively. Note that we have two error bounds ϵ_x and ϵ_y for the x and y dimensions, obtained from the training process, normally using the maximum prediction error of the function. The real cell ID of p_i is then determined by a model-biased search in the range $[C_{pred,x} - \epsilon_x, C_{pred,x} + \epsilon_x) \times [C_{pred,y} - \epsilon_y, C_{pred,y} + \epsilon_y)$.

4.3 Storage Overhead

The storage overhead is calculated by comparing the size of the indexed dataset to that of the raw dataset, which consists solely of the spatial points. The raw dataset size is 1.6 GB, while the size of the indexed dataset using DGLSI is 1.7 GB. Consequently, the storage overhead is less than 6.5%. The predominant portion of this overhead stems from the metadata required to locate the data groups in the storage system, i.e., the `manifest` file. Additionally, space is consumed by the learned RMI models in each dimension for DGLSI. Conversely, for SPRIG, the size of the "learned model" can be disregarded, as the bilinear interpolation function can be represented by four floating-point numbers.

4.4 Query Performance

Query Workload and Data Distribution. In this study, we presume the query workload is known in advance so that the index structure can be optimized. We measure the following parameters of query workloads:

1. *Nearest neighbour query workload*: the average radius R_p (in meters) of all the nearest neighbour queries.
2. *Range query workload*: the average query area R_r (measured in meters), of all range queries, calculated as the geometric mean of the width and height of the query rectangle.

Queries are randomly generated based on the parameters R_p and R_r. For nearest neighbour queries, query points are randomly generated in the service area, and the radius of the query circle follows a Gaussian distribution with mean

R_p and standard deviation $0.1 \times R_p$. For range queries, the width and height of the query rectangle follow a Gaussian distribution with mean R_r and standard deviation of $0.5 \times R_r$.

Results. Table 1 illustrates the query performance of DGLSI and SPRIG on the Geolife dataset under different query workloads. The query performance is measured by the average query execution time of 100,000 queries, and the unit is milliseconds. The measured query execution time of DGLSI and SPRIG are denoted as T_d and T_s, respectively.

Table 1. Query performance of DGLSI and SPRIG

(a) Nearest neighbor query

R_p	$\mathbf{T_d}$	T_s
200	3.77	3.89
400	8.16	8.52
800	13.24	14.63
1600	19.65	21.80

(b) Range query

R_p	$\mathbf{T_d}$	T_s
200	15.43	17.31
400	50.26	58.19
800	173.98	183.73
1600	549.42	568.64

We saw slightly better query performance of DGLSI compared to SPRIG in both nearest neighbour and range queries. One of the shortcomings of SPRIG is that the per-dimension error bounds ϵ_x and ϵ_y can be large. This is because the row IDs of the cells, as shown in Fig. 2 and Eq. (1), may change significantly along the y-axis while the encoded cell ID is monotonically increasing. This phenomenon prevents the error bounds from effectively pruning the search space. On the other hand, DGLSI uses per-dimension learned indices to predict the cell ID, which can fit more complex data distributions.

5 Conclusions and Future Work

This paper introduces DGLSI, a learned spatial index structure designed for clustered spatial data. The index structure is built on an adaptive grid-based partitioning strategy that takes into account the data distribution. We evaluated the query performance of DGLSI using the Geolife dataset and found that it outperforms SPRIG, a similar grid-based index, in nearest neighbour and range queries. Additionally, the storage overhead of DGLSI is less than 6.5% compared to the raw dataset. While learned index structures have shown significant potential in predicting the storage location of data records, we also acknowledge that the performance of DGLSI and similar spatial index structures heavily depends on the data partitioning, file organization and caching strategies employed. Moreover, the use of learned indexes slightly improves query performance compared to traditional tree-based searches but disables write operations. Updatable learned indexes should be further explored by, for example, develop an incremental algorithm to split the cells upon data insertion. In future research, we intend to delve into more sophisticated learned index structures that can dynamically and effectively partition data, even with data insertion and deletion capabilities. Furthermore, we aim to study the influence of combining file caching strategies with the learned model.

Disclosure of Interests. The authors have no competing interests to declare that are relevant to the content of this article.

References

1. Asano, T., Ranjan, D., Roos, T., Welzl, E., Widmayer, P.: Space-filling curves and their use in the design of geometric data structures. Theoret. Comput. Sci. **181**(1), 3–15 (1997). https://doi.org/10.1016/S0304-3975(96)00259-9
2. Boberg, C., Nerbonne, J., Watt, D.: The Handbook of Dialectology. Wiley, Hoboken (2018)
3. Dai, H.K., Su, H.C.: On the locality properties of space-filling curves. In: Ibaraki, T., Katoh, N., Ono, H. (eds.) ISAAC 2003. LNCS, vol. 2906, pp. 385–394. Springer, Heidelberg (2003). https://doi.org/10.1007/978-3-540-24587-2_40
4. Guttman, A.: R-trees: a dynamic index structure for spatial searching. ACM SIGMOD Rec. **14**(2), 47–57 (1984). https://doi.org/10.1145/971697.602266
5. Kraska, T., Beutel, A., Chi, E.H., Dean, J., Polyzotis, N.: The case for learned index structures. In: Proceedings of the 2018 International Conference on Management of Data, pp. 489–504. SIGMOD 2018, Association for Computing Machinery, New York, NY, USA (2018). https://doi.org/10.1145/3183713.3196909
6. Land, A.H., Doig, A.G.: An automatic method of solving discrete programming problems. Econometrica **28**(3), 497–520 (1960). https://doi.org/10.2307/1910129
7. Qi, J., Liu, G., Jensen, C.S., Kulik, L.: Effectively learning spatial indices. Proc. VLDB Endow. **13**(12), 2341–2354 (2020). https://doi.org/10.14778/3407790.3407829
8. Tan, Q., Xu, X.: Comparative analysis of spatial interpolation methods: an experimental study. Sens. Transducers **165**(2), 155 (2014)
9. Wang, H., Fu, X., Xu, J., Lu, H.: Learned Index for Spatial Queries. In: 2019 20th IEEE International Conference on Mobile Data Management (MDM), pp. 569–574 (2019). https://doi.org/10.1109/MDM.2019.00121
10. Zhang, S., Ray, S., Lu, R., Zheng, Y.: SPRIG: a learned spatial index for range and kNN queries. In: 17th International Symposium on Spatial and Temporal Databases, pp. 96–105. SSTD 2021, Association for Computing Machinery, New York, NY, USA (2021). https://doi.org/10.1145/3469830.3470892
11. Zhang, S., Ray, S., Lu, R., Zheng, Y.: Efficient learned spatial index with interpolation function based learned model. IEEE Trans. Big Data **9**(2), 733–745 (2023). https://doi.org/10.1109/TBDATA.2022.3186857
12. Zou, B., Zeng, M., Zhu, C., Xiao, L., Chen, Z.: A learned prefix bloom filter for spatial data. In: Strauss, C., Cuzzocrea, A., Kotsis, G., Tjoa, A.M., Khalil, I. (eds.) DEXA 2022. LNCS, vol. 13426, pp. 336–350. Springer, Cham (2022). https://doi.org/10.1007/978-3-031-12423-5_26

Advanced Architectures

Coordination-Free Replicated Datalog Streams with Application-Specific Availability

Owais Qayyum(✉) and Weihai Yu

UiT The Arctic University of Norway, Tromsø, Norway
{o.qayyum,weihai.yu}@uit.no

Abstract. Data are continuously generated, processed and consumed everywhere. However, the computing devices may be disconnected from the network. We want the disconnection to have as little impact on the availability of data as possible. This paper presents an approach to transform a sequential Datalog program into a distributed one. The distributed program meets the specified availability and fault-tolerance requirements. The continuous execution of the distributed program is coordination free and eventual consistent. To achieve this, we augment the Datalog program with a composite CRDT (Conflict-free Replicated Data Type). We have implemented a prototype and run experiments to study the performance characteristics of the approach.

Keywords: Recursive dataflow · CRDT · data provenance · semiring · parallel correctness · eventual consistency · fault-tolerance

1 Introduction

We generate, process and consume data over a continuum of computing devices, ranging from the edge to the cloud. The devices may be offline, but we still expect the data to be available, whenever and wherever they are needed.

There is already wonderful work on highly available distributed data systems. On the one hand, local-first software [15] with eventual-consistent data replication allows data to be always available on local devices [22,24]. On the other hand, large-scale distributed programs manage to process and deliver data at due places [13,16,23,27]. However, the current state-of-the-art is still limited. More specifically, the former is restricted to replication of base (i.e. not derived) data, while the latter can only tolerate network partitioning with expensive coordination mechanisms.

We present PRAD (Partitioned, Replicated, Asynchronous Datalog), a Datalog extension focusing on data availability and tolerance to network partitioning.

The contributions of this paper are the following.

- We propose PRAD, a Datalog extended with site configurations that specify where the input and output data should (always) be available.
- Given a PRAD program, we generate plans for continuous distributed executions that meet the specified availability and additional fault-tolerance requirements. The executions are coordination free and eventual consistent. Network messages are not required to be ordered or duplication free.

J. Tekli et al. (Eds.): ADBIS 2024, LNCS 14918, pp. 155–169, 2024.
https://doi.org/10.1007/978-3-031-70626-4_11

- To achieve eventual-consistent replication of derived data, we augment a Datalog program with a CRDT (Conflict-free Replicated Data Type), composed of a semiring provenance [8] and a causal-length set [26]. The augmented program produces the same result as the original sequential Datalog program and terminates to the same fixpoint. The input can be both insertion and deletion of facts.

The paper is organized as the following. Section 2 gives an overview of PRAD with an example and highlights the challenge. Section 3 presents preliminaries, including some common terminologies and the research work on which this paper is built. Sections 4 and 5 present our main contributions. Due to space limit, we omit the proof of the theorems, most of which are straightforward. Section 6 presents some experimental results for performance study. Section 7 discusses related work. Section 8 concludes.

2 PRAD Overview

We use the classical transitive-closure example to describe the main features of PRAD. The example Datalog program, *Ps*, consists of rules defined on relations. Relation $\mathsf{L}(\mathsf{From}, \mathsf{To})$ contains direct links between cities. $\mathsf{P}(\mathsf{From}, \mathsf{To})$ contains paths connected through the links. $\mathsf{CityFoo}(\mathsf{City})$ contains cities which Foo travels from or to, and $\mathsf{PFoo}(\mathsf{From}, \mathsf{To})$ contains the paths from or to these cities. Given the links and Foo's cities as input, Rules (1) and (2) generate all paths, and Rules (3) and (4) generate all paths specific to Foo.

$$\mathsf{P}(X, Y) \leftarrow \mathsf{L}(X, Y) \tag{1}$$

$$\mathsf{P}(X, Y) \leftarrow \mathsf{L}(X, Z), \mathsf{P}(Z, Y) \tag{2}$$

$$\mathsf{PFoo}(X, Y) \leftarrow \mathsf{CityFoo}(X), \mathsf{P}(X, Y) \tag{3}$$

$$\mathsf{PFoo}(X, Y) \leftarrow \mathsf{CityFoo}(Y), \mathsf{P}(X, Y) \tag{4}$$

Figure 1 shows an example of L facts. Program *Ps* derives P facts $\mathsf{P}(1, 3)$, $\mathsf{P}(1, 4)$ etc. Given fact $\mathsf{CityFoo}(2)$, *Ps* derives output facts $\mathsf{PFoo}(1, 2)$, $\mathsf{PFoo}(2, 2)$, $\mathsf{PFoo}(2, 3)$ and $\mathsf{PFoo}(2, 4)$.

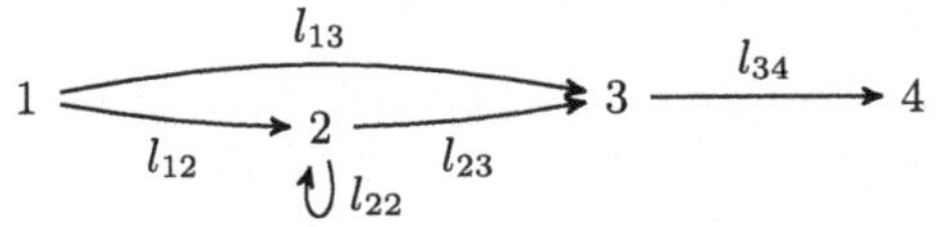

Fig. 1. Example L facts

PRAD extends a normal Datalog program with relation $\mathsf{Cfg}(\mathsf{RelName}, \mathsf{AttrName}, \mathsf{AttrVal}, \mathsf{Site})$ for site configurations. Site is a domain of network sites. Suppose we have the relations SiteEU and SiteUS for EU and US sites, and

SiteFoo for Foo's computing devices. Suppose also that relations CityEU(City) and CityUS(City) contain EU and US cities. Below are some site-configuration rules:

$$\mathsf{Cfg}(\text{“L”}, \text{“To”}, X, Y) \leftarrow \mathsf{CityEU}(X), \mathsf{SiteEU}(Y) \tag{5}$$
$$\mathsf{Cfg}(\text{“L”}, \text{“To”}, X, Y) \leftarrow \mathsf{CityUS}(X), \mathsf{SiteUS}(Y) \tag{6}$$
$$\mathsf{Cfg}(\text{“PFoo”}, *, *, Y) \leftarrow \mathsf{SiteFoo}(Y) \tag{7}$$

Rule (5)/(6) states that input of links *to* EU/US cities should be available at all EU/US network sites. Rule (7) states that output of Foo's paths should be available on all Foo's devices. Here, * is a special value, meaning "*any*". We should be able to insert and delete links at any available EU/US sites, even when the sites are offline. Moreover, Foo's paths as output result should be updated as long as the program is running and Foo's devices are online. When a device is offline, the result before it became offline should still be accessible.

One particular challenge is that an output fact may have multiple derivations (e.g. multiple paths between two cities) and the same derivation may be derived multiple times due to replication of input and intermediate data. It is important that when new facts are inserted and old facts deleted, all devices eventually receive the same output results.

System Model. Network sites may crash. Some crashed sites may restart to a previous state. Network messages may duplicate and be reordered, but they will eventually be delivered.

3 Preliminaries

This section presents the terminologies and existing work which our work is built on. When presenting the concepts, we also describe how they relate to our work.

3.1 Datalog, Distribution Policy and Parallel Correctness

Let **var** be a domain of variables, disjoint from a domain of values **dom**. A Datalog *schema* is a set of relation names. An *atom* $R(\mathbf{x})$ consists of a relation name R and a tuple of terms $\mathbf{x} = (x_1, \ldots, x_k)$. A *term* x_i is either a variable from **var** or a value from **dom**. A *fact* is an atom where every term is a value. A *universe* $\mathcal{U}$ is a set of all possible facts. A Datalog *rule* τ, $R(\mathbf{x}) \leftarrow R_1(\mathbf{x_1}), \ldots, R_n(\mathbf{x_n})$, consists of a *head*, $head_\tau = R(\mathbf{x})$, and a *body*, $body_\tau = \{R_1(\mathbf{x_1}), \ldots, R(\mathbf{x_n})\}$, where every variable in $\mathbf{x}$ must occur in some $\mathbf{x_i}$. A *Datalog program*, or just *program*, is a finite set of Datalog rules.

For a program P, $edb(P)$, the schema for *extensional database*, is the set of relation names that do not occur in the heads of P, and $idb(P)$, the schema for *intentional database*, is the set of relation names that occur in the heads. $sch(P) = edb(P) \cup idb(P)$ is the schema of P. Program P also has an *output schema*, $out(P) \subseteq sch(P)$. For the example program Ps in Sect. 2, $edb(Ps) = \{\mathsf{L}, \mathsf{CityFoo}\}$, $idb(Ps) = \{\mathsf{P}, \mathsf{PFoo}\}$, and $out(Ps) = \{\mathsf{PFoo}\}$.

An *instance* in a schema is a set of facts in the schema. Let I be an instance in schema σ and a sub-schema $\sigma' \subseteq \sigma$, we write $I_{|\sigma'}$ as the instance in σ'.

A *valuation* for rule τ in instance I is a mapping v_τ from the terms in τ to values in **dom** that preserves the facts in I. A valuation v_τ *derives* fact $v_\tau(head_\tau)$ over instance I if $v_\tau(body_\tau) \subseteq I$.

The *immediate consequence operator* for program P, T_P, applies all rules in P exactly once over the current instance I and updates I with the derivations. For input I_0 over $edb(P)$, the computational semantics of P over I_0, $P(I_0)$, is the fixpoint reached after iterative applications of T_P, starting with I_0. The result of the program is $P(I_0)_{|out(P)}$.

A *network* is a finite non-empty set $S \subseteq \mathbf{dom}$ of values representing *sites*. A *distribution policy* D over schema σ and network S is a function $\mathsf{facts}_D\colon S \rightarrow \mathcal{P}(\mathcal{U}_{|\sigma})$ mapping sites to subsets of facts in σ. For database instance I and site i, $I_{|i}$ is the local instance at i, i.e. $I_{|i} = \mathsf{facts}_D(i) \cap I$. A site i is *responsible* for fact f if $f \in \mathsf{facts}_D(i)$. $\mathsf{sites}_D(f)$ is the set of sites responsible for f. We may omit the subscript D and simply write facts and sites when the distribution policy is clear from the context.

A *PRAD site configuration* of program P (Sect. 2) is a distribution policy C over $edb(P) \cup out(P)$. A *PRAD execution policy* of P is a distribution policy E over $sch(P)$ that subsumes C.

A distributed execution consists of two parts: local execution and messaging. Every site i applies the rules with instance $I_{|i}$, and the derived facts are sent to their responsible sites, asynchronously. A distribution policy *supports* a valuation v_τ in I if there is a site i and a valuation v_τ in $I_{|i}$. A policy *strongly supports* a program P if it supports all valuations in P. A site reaches its local fixpoint when no new facts are derived locally. A program reaches a global fixpoint when all sites reach local fixpoints.

For program P, we write $[P, D](I)$ as the union of facts at all sites at global fixpoint under distribution policy D. Program P is *parallel correct* under distribution policy D if for any instance I, $[P, D](I)_{|out(P)} = P(I)_{|out(P)}$. Strong support is sufficient for parallel correctness [14].

3.2 CRDT

A CRDT [22] is an abstraction specifically designed for data asynchronously replicated at different sites. A site queries and updates its local replica without coordination with other sites. From time to time, the sites send their updates asynchronously to other sites, and merge the received updates with their local data, without any coordination with other sites. A CRDT guarantees that when all sites have applied the same set of updates, their states converge.

In a *state-based* CRDT [22], data states are partially ordered and form a join-semilattice, and state updates must be inflationary (i.e., the new state must be grater than or equal to the old one). A *join-semilattice* is a partially ordered set where every subset has a LUB (Least Upper Bound). LUB is also known as a *join* and is written as $\sqcup$[1]. The merge of two CRDT states s_1 and s_2 is the result

[1] To avoid being confused with the join $\bowtie$ of relations, in the rest of the paper, we use the term *merge* for $\sqcup$.

of $s_1 \sqcup s_2$. With *delta-state* CRDTs [2,7], it is sufficient to only send and merge *join-irreducible* states (i.e. the updated deltas instead of the entire states).

We adopt two specific CRDTs, namely p-semiring CRDT and CLSet (Causal-Length Set [26]) CRDT. A *p-semiring* ("p" stands for provenance) [21] is a commutative semiring with additional properties. More specifically, a p-semiring is an algebraic structure $\langle E, +, \cdot, \mathbf{0}, \mathbf{1}\rangle$ such that: i) E is a set and $\mathbf{0}, \mathbf{1} \in E$; ii) $\langle E, +, \mathbf{0}\rangle$ is a commutative monoid, i.e., $\mathbf{0}$ is the unit element, $\mathbf{0}+e = e+\mathbf{0} = e$, and $+$ is both associative and commutative; iii) $\langle E, \cdot, \mathbf{1}\rangle$ is also a commutative monoid; iv) $\mathbf{0}$ is an absorbing element of $\cdot$, i.e., $\mathbf{0} \cdot e = e \cdot \mathbf{0} = \mathbf{0}$; v) $\cdot$ left- and right-distributes over $+$; vi) $+$ and $\cdot$ are idempotent, i.e., $e + e = e$ and $e \cdot e = e$; and vii) $+$ applies to empty and infinitive sets, i.e., $\sum \emptyset = \mathbf{0}$ and $\sum E = \mathbf{1}$.

As a common convention, we may write $e_1 \cdot e_2$ as $e_1 e_2$.

The properties of $+$ allow us to define a partial order $\leq_p$: $e_1 \leq_p e_2$ iff $e_1 + e_2 = e_2$. This partial order has the properties $\mathbf{0} \leq_p e_1e_2 \leq_p e_1 \leq_p e_1 + e_2 \leq_p \mathbf{1}$. As the maximum element in E, $\mathbf{1}$ is an absorbing element of $+$. That is, $\mathbf{1} + e = e + \mathbf{1} = \mathbf{1}$. In fact, $\langle E, \leq_p\rangle$ is a distributive lattice [5] (thus also a join-semilettice), where $+$ coincides with $\sqcup$, $\cdot$ coincides with $\sqcap$ (Greatest Lower Bound), and $+$ distributes over $\cdot$, i.e., $e_1 + e_2e_3 = (e_1 + e_2)(e_1 + e_3)$. Finally, there is a useful absorption property: $e_1 + e_1e_2 = e_1$.

CLSet [26] is a CRDT for set, where each element is associated with a *causal length*. Intuitively, insertion and deletion of an element are inverse operations of one another. Each time the membership of an element changes, its causal length increments by one, which reverses the effect of a previous update. As shown in Fig. 2, the state of a CLSet is a function that associates every element with a causal length (initially 0). ins^Δ and del^Δ are delta-mutators that return join-irreducible states instead of the entire states. An element e is regarded as being in the set when its causal length is an odd number. A local effective insertion or deletion simply increments the causal length of the element by one, which is obviously inflationary. After the merge of two sets cl and cl', the new causal length of element e is the maximum of the causal lengths of e in cl and cl'.

$$\mathsf{CLSet}(E) \stackrel{\text{def}}{=} cl\colon E \to \mathbb{N}_0 \quad \text{initially } \forall e \in E\colon cl(e) = 0$$

$$\mathsf{in?}(cl, e) \stackrel{\text{def}}{=} \mathsf{odd?}(cl(e))$$

$$(cl \sqcup cl')(e) \stackrel{\text{def}}{=} \mathsf{max}(cl(e), cl'(e))$$

$$\mathsf{ins}^\Delta(cl, e) \stackrel{\text{def}}{=} \begin{cases} \{e \mapsto cl(e) + 1\} & \text{if } \neg\mathsf{in?}(cl, e) \\ \{\} & \text{otherwise} \end{cases}$$

$$\mathsf{del}^\Delta(cl, e) \stackrel{\text{def}}{=} \begin{cases} \{e \mapsto cl(e) + 1\} & \text{if } \mathsf{in?}(cl, e) \\ \{\} & \text{otherwise} \end{cases}$$

Fig. 2. CLSet CRDT [26]

4 CRDT-Augmented Datalog

A Datalog^{p}program is a Datalog program augmented with p-semiring and causal lengths (Sect. 3.2). Below, we transform a normal Dalalog program P into a Datalog^p program $\ddot{P}$.

Let **prv** be a domain of provenance expressions in p-semiring, disjoint from **var** and **dom**. A *p-instance*, $\ddot{I}: \mathcal{U} \rightarrow \mathbf{prv}$, is a function from the univers of facts $\mathcal{U}$ to **prv**. A p-instance $\ddot{I}$ is derived from an instance I of a normal Datalog program. For fact $f \in I$, the pair $f \mapsto \ddot{I}(f)$ is a *p-fact* in $\ddot{I}$. The *support* of a p-instance is the set of facts whose provenance expressions are not $\mathbf{0}$, the smallest value in $\langle \mathbf{prv}, \leq_p \rangle$. That is, $\mathbf{supp}(\ddot{I}) = \{f \in \mathcal{U} \mid \ddot{I}(f) \neq \mathbf{0}\}$.

We define an order $\ddot{\leq}$ over p-instances. For two p-instances $\ddot{I}_1$ and $\ddot{I}_2$, $\ddot{I}_1 \ddot{\leq} \ddot{I}_2$ if and only if $\forall f \in \mathcal{U}\colon \ddot{I}_1(f) \leq_p \ddot{I}_2(f)$. Clearly, if $\ddot{I}_1 \ddot{\leq} \ddot{I}_2$, then $\mathbf{supp}(\ddot{I}_1) \subseteq \mathbf{supp}(\ddot{I}_2)$.

Next, we describe how p-facts are derived, using program *Ps* in Sect. 2 and the example instance in Fig. 1.

Upon the insertion of an edb fact f, p-fact $f \mapsto id_f$ is inserted in $\ddot{I}$. That is

$$\ddot{I}(f) = id_f \tag{8}$$

where $id_f \in \mathbf{uid} \subset \mathbf{prv}$ is a unique identifier of f. For the example instance in Fig. 1, suppose $id_{\mathsf{L}(1,2)} = l_{12}$, then, $\ddot{I}(\mathsf{L}(1,2)) = f_{12}$ and $\mathsf{L}(1,2) \mapsto l_{12}$ is a p-fact.

For a Datalog valuation v of rule τ in instance I, there is a Dadalog^p valuation $\ddot{v}$ in the corresponding p-instance $\ddot{I}$, such that

$$\ddot{I}(v(head_\tau)) = \prod_{f \in v(body_\tau)} \ddot{I}(f) \tag{9}$$

For example, if p-instance $\ddot{I}$ contains two p-facts $\mathsf{L}(1,2) \mapsto l_{12}$ and $\mathsf{P}(2,3) \mapsto l_{23}$ (Fig. 1), then a valuation of Rule (2) in $\ddot{I}$ derives p-fact $\mathsf{P}(1,3) \mapsto l_{12}l_{23}$.

The union of two p-instances $\ddot{I}_1$ and $\ddot{I}_2$ is defined as

$$\forall f \in \mathcal{U}\colon (\ddot{I}_1 \cup \ddot{I}_2)(f) = \ddot{I}_1(f) + \ddot{I}_2(f) \tag{10}$$

For example, if p-instance $\ddot{I}_1$ contains p-fact $\mathsf{P}(1,3) \mapsto l_{13}$ and $\ddot{I}_2$ contains $\mathsf{P}(1,3) \mapsto l_{12} \cdot l_{23}$ (Fig. 1), then $\ddot{I}_1 \cup \ddot{I}_2$ derives p-fact $\mathsf{P}(1,3) \mapsto l_{13} + l_{12} \cdot l_{23}$.

Notice that the union of p-instances coincides with $\sqcup$ of p-instances under $\ddot{\leq}$. That is, $\ddot{I}_1 \cup \ddot{I}_2 = \ddot{I}_1 \sqcup \ddot{I}_2$. Just like the union of normal Datalog instances being inflationary under $\subseteq$, the union of p-instances is inflationary under $\ddot{\leq}$.

Clearly, $\mathbf{supp}(\ddot{I}_1 \cup \ddot{I}_2) = \mathbf{supp}(\ddot{I}_1) \cup \mathbf{supp}(\ddot{I}_2)$.

The immediate consequence operator for program $\ddot{P}$ is

$$T_{\ddot{P}}(\ddot{I}) = \ddot{I} \cup \bigcup_{\tau \in P, v_\tau(body_\tau) \subseteq I} \{v_\tau(head_\tau) \mapsto \prod_{f \in v_\tau(body_\tau)} \ddot{I}(f)\} \tag{11}$$

Since $T_{\ddot{P}}$ is inflationary and the $\cdot$ and $+$ operations are idempotent, for any p-instance $\ddot{I}$ with finite support, the iterative application of $T_{\ddot{P}}$, starting with $\ddot{I}$, reaches a fixpoint. We write the p-instance at the fixpoint as $\ddot{P}(\ddot{I})$.

Theorem 1. *For a Datalog program P and instance I, and the corresponding Datalog^p program $\ddot{P}$ and p-instance $\ddot{I}$ obtained with Equations (8)– (11),* $\mathbf{supp}(\ddot{P}(\ddot{I})) = P(I)$.

Table 1. Applying $T_{\ddot{Ps}}$

$\ddot{I}_0$	$\{\mathsf{L}(1,2) \mapsto l_{12}, \mathsf{L}(1,3) \mapsto l_{13}, \mathsf{L}(2,2) \mapsto l_{22}, \mathsf{L}(2,3) \mapsto l_{23}, \mathsf{L}(3,4) \mapsto l_{34}\}$
$\ddot{I}_1^{\Delta}$	$\{\mathsf{P}(1,2) \mapsto l_{12}, \mathsf{P}(1,3) \mapsto l_{13}, \mathsf{P}(2,2) \mapsto l_{22}, \mathsf{P}(2,3) \mapsto l_{23}, \mathsf{P}(3,4) \mapsto l_{34}\}$
$\ddot{I}_2^{\Delta}$	$\{\mathsf{P}(1,3) \mapsto l_{12}l_{23}, \mathsf{P}(1,4) \mapsto l_{13}l_{34}, \mathsf{P}(2,4) \mapsto l_{23}l_{34}\}$
$\ddot{I}_3^{\Delta}$	$\{\mathsf{P}(1,4) \mapsto l_{12}l_{23}l_{34}\}$

Table 1 shows the iterative applications of the immediate consequence operator $T_{\ddot{Ps}}$ to program $\ddot{Ps}$, with the initial p-instance $\ddot{I}_0$ as shown in Fig. 1.

The valuations of Rule (1) of program $\ddot{Ps}$ in $\ddot{I}_0$ derive new P facts in p-instance $\ddot{I}_1$. Table 1 only shows the new derived facts in $\ddot{I}_1^{\Delta}$. That is, $T_{\ddot{P}}(\ddot{I}_0) = \ddot{I}_0 \cup \ddot{I}_1^{\Delta} = \ddot{I}_1$ and $\ddot{I}_0 \cap \ddot{I}_1^{\Delta} = \emptyset$.

The valuations of Rule (2) in $\ddot{I}_1$ derive new p-facts in $\ddot{I}_2^{\Delta}$. Notice that the new derivations for $\mathsf{P}(1,2) \mapsto l_{12}l_{22}$ and $\mathsf{P}(2,3) \mapsto l_{22}l_{23}$ are not visible in $\ddot{I}_2^{\Delta}$. In fact, due to the absorption property (Sect. 3.2), $\ddot{I}_2(\mathsf{P}(1,2)) = l_{12} + l_{12}l_{22} = l_{12} = \ddot{I}_1(\mathsf{P}(1,2))$. Similarly, further applications of $T_{\ddot{Ps}}$ on $\ddot{I}_3$ do not lead to any new derivation. In fact, the absorption property prevents endless inflationary updates due to loops. Thus $\ddot{I}_3$ is a fixpoint of $T_{\ddot{Ps}}$ and $\ddot{Ps}(\ddot{I}_0) = \ddot{I}_3 = \ddot{I}_0 \cup \ddot{I}_1^{\Delta} \cup \ddot{I}_2^{\Delta} \cup \ddot{I}_3^{\Delta}$.

The p-semiring provenance of an idb fact f describes how f depends on edb facts. We can write p-semiring expressions in canonical normal form (CNF), as sum of products. For $\mathsf{P}(1,4)$, the provenance expression $l_{13}l_{34} + l_{12}l_{23}l_{34}$ is in CNF while the equivalent expression $(l_{13} + l_{12}l_{23})l_{34}$ is not. Notice that the two sets, $\{\mathsf{L}(1,3), \mathsf{L}(3,4)\}$ and $\{\mathsf{L}(1,2), \mathsf{L}(2,3), \mathsf{L}(3,4)\}$, corresponding to the two products in CNF, are exactly the two smallest derivations of $\mathsf{P}(1,4)$. The following proposition describes how idb facts depend on edb facts.

Proposition 1. *Given Datalog program P, instance I, and the corresponding Datalog^p program $\ddot{P}$ and p-instance $\ddot{I}$. Given idb fact $f \in I$, $\ddot{I}(f) = \sum_i (\prod_j e_{i,j})$. For any proof tree T of f, there is an i, such that the set of $e_{i,j}$ elements are exactly the set of the identifiers of the edb facts at the leave nodes of T.*

A Datalog^p program $\ddot{P}$ in a distribution policy is *eventual consistent* if it is both parallel correct (Sect. 3.1) and eventually convergent. A p-instance $\ddot{I}$ is *convergent* if for any fact f and $i, j \in \mathsf{sites}(f)$, $\ddot{I}_{|i}(f) = \ddot{I}_{|j}(f)$. That is, the provenance expression of the fact is the same at all sites that are responsible for the fact. The program $\ddot{P}$ is *convergent* if starting with any convergent p-instance $\ddot{I}$, when $\ddot{P}$ reaches a global fixpoint, $T_{\ddot{P}}(\ddot{I})$ is convergent.

A distributed execution is coordination-free when the sites in the network do not exchange messages beyond edb or idb facts. Examples of coordination include acknowledgment of message delivery and protocols for achieving some consensus [3]. For a Datalog program that continuously applies valuations with the semi-naïve algorithm [1], a distributed execution being coordination-free effectively means that valuations of rules at different sites can be applied multiple times and in different order, and the execution still generates the same result as a sequential one.

It is known that for any inflationary Datalog program, rules can be repeatedly applied in any order [3]. This result applies also to Datalog^p programs.

Proposition 2. *Given a Datalog program P and instance I, and the corresponding Datalog^p program $\ddot{P}$ and p-instance $\ddot{I}$. For any distribution policy D that strongly supports $\ddot{P}$, coordination-free executions end up with the same result* $\mathbf{supp}([\ddot{P}, D](\ddot{I})) = P(I)$.

Since $\langle \mathbf{prv}, \leq_p \rangle$ (and thus $\langle \ddot{I}, \ddot{\leq} \rangle$) is a distributive lattice, p-instances of the same relation at different sites in fact form a state-based CRDT. They are therefore eventually convergent without any explicit coordination.

Datalog^p achieves what Datalog is already capable of, that is, coordination-free and eventual-consistent maintenance of inflationary program streams. The real challenge is to handle deletion of edb facts, since deletion is not an inflationary update. What Datalog^p with p-semiring provenance provides, is how idb facts are dependent on edb facts. Equipped with this knowledge, we now apply CLSet CRDT (Sect. 3.2) to support the coordination-free execution of Datalog programs with both insertion and deletion operations.

We first introduce a causal-length function: $\mathsf{cl}\colon \mathbf{uid} \to \mathbb{N}_0$.

For an edb fact f, the causal length of its identifier id_f is updated as shown in Fig. 2. As a CRDT, replicas of cl are convergent without any coordination.

We then introduce an interpretation of provenance expressions over cl.

$$\begin{aligned} \llbracket\rrbracket : \mathbf{prv} \times (\mathbf{uid} \to \mathbb{N}_0) &\to \mathbb{B} \\ \llbracket id \rrbracket_{\mathsf{cl}} &= \mathsf{odd?}(\mathsf{cl}(id)) \\ \llbracket e_1 + e_2 \rrbracket_{\mathsf{cl}} &= \llbracket e_1 \rrbracket_{\mathsf{cl}} \vee \llbracket e_2 \rrbracket_{\mathsf{cl}} \\ \llbracket e_1 \cdot e_2 \rrbracket_{\mathsf{cl}} &= \llbracket e_1 \rrbracket_{\mathsf{cl}} \wedge \llbracket e_2 \rrbracket_{\mathsf{cl}} \end{aligned}$$

A fact is in the database if its provenance expression interprets to be True. For $l \in \{l_{12}, l_{13}, \ldots\}$ in Table 1, initially $\mathsf{cl}(l) = 1$. If we delete $\mathsf{L}(1,2)$, $\mathsf{cl}'(l_{12}) = 2$. $\llbracket \ddot{I}_3(\mathsf{P}(1,3)) \rrbracket_{\mathsf{cl}'} = \llbracket l_{13} + l_{12} l_{23} \rrbracket_{\mathsf{cl}'} = \llbracket l_{13} \rrbracket_{\mathsf{cl}'} \vee \llbracket l_{12} l_{23} \rrbracket_{\mathsf{cl}'} = \mathrm{True} \vee (\llbracket l_{12} \rrbracket_{\mathsf{cl}'} \wedge \llbracket l_{23} \rrbracket_{\mathsf{cl}'}) =$ True $\vee$ (False $\wedge$ True) = True. Therefore, $\mathsf{P}(1,3)$ remains in the database.

Lemma 1. *Given a Datalog program P and instance I, and the corresponding Datalog^p program $\ddot{P}$ and p-instance $\ddot{I}$.* $\llbracket \ddot{P}(\ddot{I}) \rrbracket_{\mathsf{cl}}(f) =$ *True if and only if* $f \in P(I)$.

Theorem 2. *Given a Datalog program P and instance I, and the corresponding Datalog^p program $\ddot{P}$ and p-instance $\ddot{I}$. For any distribution policy D that strongly supports $\ddot{P}$, for any coordination-free execution of $\ddot{P}$,* $\llbracket [\ddot{P}, D](\ddot{I}) \rrbracket_{\mathsf{cl}}(f) =$ *True if and only if* $f \in P(I)$.

5 Coordination-Free Program Execution

A PRAD program specifies where input and output facts should be stored. An execution of the program aims at making the facts available at the specified sites as soon as possible.

The PRAD runtime is essentially an instance of a running Datalog^p program where every relation $\ddot{R}$ has a provenance attribute added to the Datalog relation R. The provenance expressions are updated according to Eqs. (8)–(11). In addition, every site maintains a cl-map for the causal lengths of the identifiers of the edb facts known to the site. Deletions of edb facts will cause the cl-map updated. When the program is running, the idb relations and the cl-map are incrementally updated as long as the sites are connected. For every output instance $\ddot{Q}$, there is also a Datalog instance Q, located at the same site of $\ddot{Q}$, which is updated with the interpretation of the provenance expressions over causal lengths (Sect. 4). That is, while the updates in $\ddot{Q}$ are inflationary, facts in Q might be deleted and thus is not inflationary.

For a bottom-up execution of a Datalog program, the valuations of a rule consist of the evaluation of relational algebra operations [1]. For example, rule $R(Y, Z) \leftarrow R_1(Y, X), R_2(X, Z)$ can be evaluated as $\pi_{Y,Z}(R_1 \bowtie_X R_2)$. An execution has *parallel degree* N_p if every join operation can be executed in parallel on at least N_p sites.

Proposition 3. *Consider a join operation $R_1 \bowtie_X R_2$. If (1) R_1 and R_2 have the same partitioning defined on attribute X, and (2) the partitions of R_1 and R_2 have the same distribution policy defined on attribute X such that the partitions are stored on N_p different sites, then, the join operation has parallel degree N_p.*

This property of join is obvious and is the foundation of most parallel join algorithms such as MapReduce [6].

An execution has *replication degree* N_r if every valuation is supported by at least N_r sites. Clearly, an alternative definition of a distribution policy strongly supporting a program (Sect. 3.1) is that the executions have replication degrees $N_r \geq 1$. An execution with replication degree N_r can tolerate $N_r - 1$ simultaneous site failures.

Proposition 4. *Consider rule τ: $R \leftarrow R_1(X, Y_1), R_2(X, Y_2)$. If (1) R_1 and R_2 have the same partitioning defined on attribute X, and (2) the partitions of R_1 and R_2 have the same distribution policy defined on attribute X such that each partition is stored on N_r different sites, then, all valuations of τ have replication degree N_r.*

Next, we present an algorithm that generates a distribution policy such that every join operation has a parallel degree non-less than N_p and every rule valuation has a replication degree non-less than N_r.

For program P, we maintain a list $\mathcal{J}$ of join attributes and process the list until it is empty. Every element in $\mathcal{J}$ is a set $J = \{(R_i, k_i) | 1 \leq i \leq n\}$, where $R_i \in sch(P)$, and $\{(R_i, k_i), (R_j, k_j)\} \subseteq J$ if there is a rule τ: $\{R_i(\mathbf{x_i}), R_j(\mathbf{x_j})\} \subseteq body_\tau$, and there is a join on the k_i-th attribute of R_i and the k_j-th attribute of R_j, or $R_i \bowtie_{\mathbf{x_i}[k_i]=\mathbf{x_j}[k_j]} R_j$.

In what follows, we write R_i and R_j to denote two different relations if $i \neq j$, and $[R]_m^k$ and $[R]_{m'}^{k'}$ to denote different relation *names* if $k \neq k'$ or $m \neq m'$, but they have the same attributes as R.

When processing element J in $\mathcal{J}$, we partition each R_i with hash function $h_J\colon \mathcal{D}_J \rightarrow 1..N_p$, where $\mathcal{D}_J$ is the common domain of attributes (R_i, k_i), and create relations $[R_i]^{k_i}_{p,r}$, where $1 \leq p \leq N_P$ and $1 \leq r \leq N_r$ for the r-th replica of the p-th partition. The distribution policy is such at $\mathsf{sites}([R_i]^{k_i}_{p,r}(\mathbf{x_i})) = \mathsf{sites}([R_j]^{k_j}_{p',r'}(\mathbf{x_j}))$ if and only if $h_J(\mathbf{x_i}[k_i]) = h_J(\mathbf{x_j}[k_j]) = p = p'$. Note that for any fact $[R_i]^{k_i}_{p,r}(\mathbf{x_i})$, $|\mathsf{sites}([R_i]^{k_i}_{p,r}(\mathbf{x_i}))| = 1$.

The algorithm further accounts for some additional situations. Due to space limit, we only present some of them briefly.

When a relation R has multiple partitioning defined on different attributes, most facts are over-replicated. For example, if relation P of the example program *Ps* is partitioned on both From and To attributes and each partition has a replication degree 2, then the facts with different From and To values are replicated at 4 different sites. To deal with this situation, we merge the created partitions and create a new one for special facts (that have the same From and To values).

A PRAD program has already specified a site configuration. Suppose in the site configuration, an edb relation R has already n_p partitions defined on a join attribute (R, k), each partition with n_r replicas. Instead of adding (R, k) into $\mathcal{J}$, we add $([R]^k_{p,r}, k)$ $(1 \leq p \leq n_p, 1 \leq r \leq n_r)$ into the list and use $\lceil N_p/n_p \rceil$ and $\lceil N_r/n_r \rceil$ as the required parallel and replication degrees.

To maintain the required parallel and replication degrees, it is not necessary to require $\mathsf{sites}([R]^k_{p,r}(\mathbf{x})) \neq \mathsf{sites}([R]^k_{p',r'}(\mathbf{x}))$ when $p \neq p'$ or $r \neq r'$, as long as there are at least N_p sites for different partitions and N_r sites for the different replicas of the same partition. Therefore the minimum number of sites to guarantee parallel degree N_p and replication degree N_r is $\max(N_p, N_r)$.

The distribution policy generated with the above algorithm can be further associated with different data *communication strategies*. In the 1-1 strategy, one partition replica of a relation joins with exactly one partition replica of another relation, and a fact derived at an upstream site is sent to exactly one downstream site. In the N-N strategy, on the other hand, all partition replicas of a relation joins with all partition replicas of another relation, and a fact derived at the upstream replica sites is sent to all downstream replica sites.

Suppose for rule $\tau\colon Q \leftarrow R, S$, we have generated partition replicas $[Q]^k_{p,r}$, $[R]^{k_1}_{p,r}$ and $[S]^{k_2}_{p,r}$, for $1 \leq p \leq N_p$, $1 \leq r \leq N_r$. With the 1-1 strategy, for every partition p, we generate N_r rules:

$$\tau_{p,r}\colon [Q]^k_{p,r} \leftarrow [R]^{k_1}_{p,r}, [S]^{k_2}_{p,r}$$

The r-th replica of the p-th partition of R^{k_1} joins with the same r-th replica of the same p-th partition of S^{k_2}. The derived facts are later dispatched to the same replica of the same partitions of Q^k.

With the N-N strategy, we first merge, for each p-th partition of Q^k (and do the same for R^{k_1} and S^{k_2}), the different replica relation names $[Q]^k_{p,r}$ $(1 \leq r \leq N_r)$, into a single partition name $[Q]^k_p$, such that for any fact $Q(\mathbf{x})$

$$\mathsf{sites}([Q]^k_p(\mathbf{x})) = \bigcup_{1 \leq r \leq N_r} \mathsf{sites}([Q]^k_{p,r}(\mathbf{x}))$$

That is, $[Q]_p^k$ represents all replicas of the p-th partition of Q^k (respectively, $[R]_p^{k_1}$ and $[S]_p^{k_2}$). So for any fact $[Q]_p^k(\mathbf{x})$, $|\mathsf{sites}([Q]_p^k(\mathbf{x}))| = N_r$. Then we generate the following rule for every partition:

$$\tau_p\colon [Q]_p^k \leftarrow [R]_p^{k_1}, [S]_p^{k_2}$$

With the new rules, all replicas of the p-th partition of R^{k_1} join with all replicas of the p-th partition of S^{k_2}. The derived facts are later dispatched to all replicas of the same p-th partition of Q^k.

The 1-1 strategy is effectively the same as running N_r instances of the program with replication degree 1, and involves less runtime overhead than the N-N strategy. The two strategies behave differently when a site fails. With the 1-1 strategy, all down-stream sites of the failed replica stop receiving updates, while with the N-N strategy, the down-stream sites continue receiving new updates as long as the number of failed replicas of the same relation partition is less than N_r.

6 Experiments

We have implemented a prototype of PRAD in Elixir[2]. We implemented network sites using Elixir GenServers[3] and data storage using ETS[4] (Erlang Term Storage) tables. A GenServer is a very lightweight Erlang process running on the Erlang virtual machine called BEAM. We added manually delays for two purposes, first to make the experiments closer to a real-life distributed environment, and second, to overcome the limitation of the shared resources on a single computer, such as CPUs and main memory. More specifically, for every message between two sites, we added a 10ms delay, and for every search in a table, we added 1ms for every 1000 rows. We ran the experiments in Linux (Ubuntu 20.04, kernel 5.4) with an 8-core i7-9700 CPU and 32 GB of 2667 MT/s main memory. Due to the manual insertion of the delays, the hardware configuration does not have a significant impact on the performance results.

We ran the classical Datalog program for calculating transitive closures (i.e. program P_S in Sect. 2). In the first two experiments, we insert 10000 L facts for every execution, which is about ten percent of the fully connected graph. We measure the latency of every new P derivation, which is the time taken from a L fact is inserted till the P fact derived from the insertion is delivered at the specified sites. For a P fact with multiple derivations, the latency of each derivation is measured.

In the first experiment, we study the impact of partition degrees on latency. Figure 3a and Fig. 3b show the histogram of the derivation latency. In order for the plots to be readable, we show the results in two figures, where the histograms

[2] https://elixir-lang.org, version 1.14, compiled with Erlang OTP 25.0.
[3] https://hexdocs.pm/elixir/1.14/GenServer.html.
[4] https://erlang.org/doc/man/ets.html.

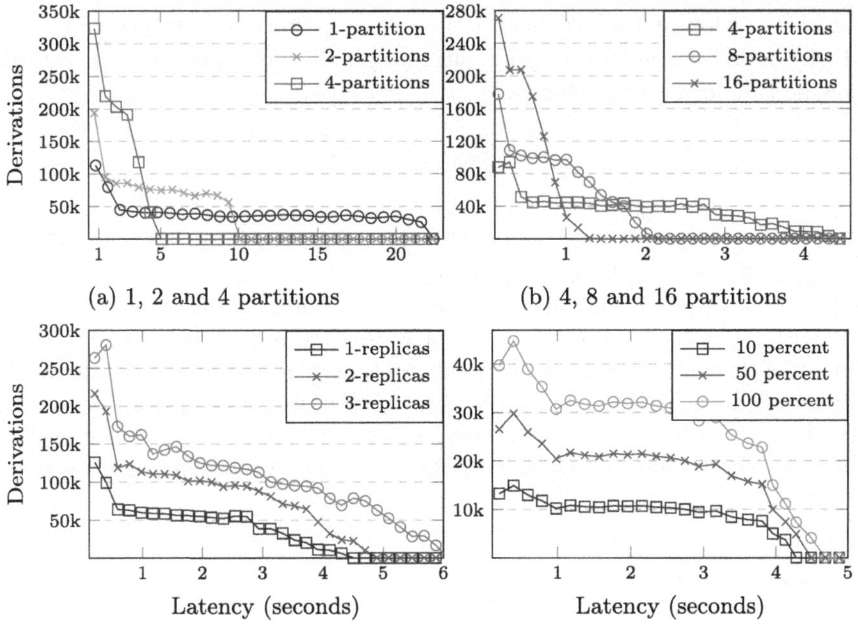

(c) 1, 2 and 3 replicas (d) Deletions

Fig. 3. Latency of derivations

for 4 partitions occur in both, with different latency ranges (as red lines connecting small boxes). The X-axis of the plots is the latency intervals. The Y-axis is the number of P derivations in each interval. As we can see, for this particular program, the executions scale well with the increasing number of partitions. The latency does not reduce any further beyond 16 partitions (not shown in the figures). The derivations with long latency are typically the ones that are derived after many iterations (i.e. paths connected with many links).

In the second experiment, we study the impact of replication on latency with the N-N communication strategy. We fix the number of partitions to 4, and vary the number of replicas. As Fig. 3c shows, when the number of replicas increases, the number of derivations with longer latency also increases. When the number of replicas increases from one to two, the latency increases up to 10%. When the number of replicas increases from one to three, the latency increases by 20%.

In the third experiment, we measure the latency of deletions, with 4 partitions and two replicas. After the insertion of 10000 L facts as well as the derivations of the P facts (that is, at the end of the execution of the second experiment), we deleted randomly 10%, 50% and 100% of the L facts. The latency is the time taken from the L facts are deleted till the P facts at the specified sites are affected (not necessarily deleted, due to alternative derivations). Deletions take a little longer than the insertions, because the executions always run on full p-instances, while the experiments for insertions start with an empty instance.

7 Related Work

Parallel execution of database operations has been widely adopted, in particular after the seminal work of MapReduce [6]. Data streaming has been supported in later systems, such as Spark [25] and Differential Dataflow [17]. Fault-tolerant approaches, either use backups to immediately replace failed sites [11], or run active replicas that produce duplicate data streams which down-stream nodes merge with [12] or switch to [4]. Both approaches require loosely synchronized clocks and reliable in-order message delivery. Research and systems for distributed Datalog have also been around for decades [13,16,20,23,27]. The focus of all these has been on data partitioning and parallel execution, typically for large-scale data processing on clusters in large data centers, where the computers in the cluster are typically supposed to be always connected. For cloud-edge settings where data may be collected and consumed outside of data centers, and the data should be available even when the devices are offline, there is a need for asynchronous data replication, in addition to partitioning.

It is known, as the CALM (Consistency And Logical Monotonicity) theorem [3,10], that when the queries are monotonous and the data input is inflationary, there is a coordination-free distributed execution of the database program that is parallel correct and eventual consistent. Therefore, earlier research on distributed Datalog works correctly for distribution policies including data replication, as long as the data input is inflationary (i.e. insertion only).

Supporting data streams, in our context, is essentially the same as incremental view maintenance, which has a long history in the relational database community. One of the key challenges is to handle deletions for recursive Datalog programs. Gupta et al. [9] presented an algorithm (later known as Deletion-Rederive, or DRed), that first deletes the idb facts derived from the deleted edb fact and then re-inserts the idb facts that have alternative derivations. Motik et al. [19] presented optimized variants of the DRed algorithm, for Datalog with stratified negation (whereas our work does not support for negation). Orchestra [13] and Declarative Networking [16,20] use data provenance to handle deletions without re-inserting idb facts. The usage of provenance is somewhat different from ours. They use the provenance to detect multiple or circular derivations, whereas we combine provenance and causal lengths that are join-semilattices and hence are state-based CRDTs desirable for handling concurrent insertions and deletions from multiple replicas.

Declarative Networking specifies site configurations of Datalog programs with an elegant notation [16]. The PRAD specification of site configuration is more general in that data not only can be partitioned, but can also be replicated. In addition, the domains of the relation attributes used for the PRAD site configuration are not restricted to network sites, which is particularly relevant for Declarative Networking whose primary application domain is networking.

Most existing CRDTs allow for eventual-consistent and coordination-free replication of base (i.e. edb in Datalog) data. Lasp [18] supports distributed coordination-free streams of functional programs. Lasp adopts OR-Set (Observed-Remove Set) CRDT [22] and supports data derived from functional programming operations, including map, reduce, filter, product, union and intersection. Most of these operations have their counterparts in relational algebra and therefore can also be applied to database relations. The Lasp metadata grows with every product (respectively join) operation, so recursive Lasp programs may not terminate.

This paper is built on our previous work [21] that uses p-semiring and causal length for incremental maintenance of replicated and partitioned relation views. In this paper, we apply these to general support of Datalog programs. In particular, given a sequential Datalog program and a specification of availability requirements of the input and out data, we are able to generate and run continuously a distributed program that meets the specified requirements as well as additional parallel and replication run-time requirements.

8 Conclusion

Existing work on distributed Datalog streams, or incremental maintenance of Datalog programs, has some limitations, when both the input and output data are required to be available on a wide continuum of computing devices, including the edge. They are either limited to partitioned but not replicated data, or they are limited to insertion-only data input. Our CRDT-based approach addresses the limitations so that data can be replicated and the input could include deletions. We have carried out some performance study, which indicates that the approach scales well with the increase of the parallel degree of the executions whereas it does incur extra runtime overhead with the increase of the replication degree for enhanced fault tolerance.

Our work has some limitations. For example, we are not able to handle negations, and the current implementation and experimentation are still rudimentary.

References

1. Abiteboul, S., Hull, R., Vianu, V.: Foundations of Databases. Addison-Wesley, Boston (1995)
2. Almeida, P.S., Shoker, A., Baquero, C.: Delta state replicated data types. J. Parallel Distrib. Comput. **111**, 162–173 (2018)
3. Ameloot, T.J., Neven, F., den Bussche, J.V.: Relational transducers for declarative networking. J. ACM **60**(2), 15:1–15:38 (2013)
4. Balazinska, M., Balakrishnan, H., Madden, S.R., Stonebraker, M.: Fault tolerance in the borealis distributed stream processing system. ACM Trans. Database Syst. **33**(1), 1–44 (2018)
5. Bistarelli, S., Montanari, U., Rossi, F.: Semiring-based constraint satisfaction and optimization. J. ACM **44**(2), 201–236 (1997)
6. Dean, J., Ghemawat, S.: Mapreduce: simplified data processing on large clusters. In: Brewer, E.A., Chen, P. (eds.) OSDI, pp. 137–150 (2004)
7. Enes, V., Almeida, P.S., Baquero, C., Leitão, J.: Efficient Synchronization of State-Based CRDTs. In: ICDE (2019)

8. Green, T.J., Karvounarakis, G., Tannen, V.: Provenance semirings. In: PODS, pp. 31–40 (2007)
9. Gupta, A., Mumick, I.S., Subrahmanian, V.S.: Maintaining views incrementally. SIGMOD Rec. **22**(2), 157–166 (1993)
10. Hellerstein, J.M.: The declarative imperative: experiences and conjectures in distributed logic. SIGMOD Rec. **39**(1), 5–19 (2010)
11. Hwang, J.H., Balazinska, M., Rasin, A., Cetintemel, U., Stonebraker, M., Zdonik, S.: High-availability algorithms for distributed stream processing. In: ICDE, pp. 779–790 (2005)
12. Hwang, J., Çetintemel, U., Zdonik, S.B.: Fast and highly-available stream processing over wide area networks. In: Alonso, G., Blakeley, J.A., Chen, A.L.P. (eds.) (ICDE), pp. 804–813 (2008)
13. Ives, Z.G., et al.: The ORCHESTRA collaborative data sharing system. SIGMOD Rec. **37**(3), 26–32 (2008)
14. Ketsman, B., Albarghouthi, A., Koutris, P.: Distribution policies for datalog. Theory Comput. Syst. **64**(5), 965–998 (2020)
15. Kleppmann, M., Wiggins, A., van Hardenberg, P., McGranaghan, M.: Local-first software: you own your data, in spite of the cloud. In: ACM Onward!, pp. 154–178 (2019)
16. Loo, B.T., et al.: Declarative networking: language, execution and optimization. In: SIGMOD, pp. 97–108 (2006)
17. McSherry, F., Murray, D.G., Isaacs, R., Isard, M.: Differential dataflow. In: CIDR (2013)
18. Meiklejohn, C., van Roy, P.: Lasp: a language for distributed, coordination-free programming. In: PPDP, pp. 184–195 (2015)
19. Motik, B., Nenov, Y., Piro, R., Horrocks, I.: Maintenance of datalog materialisations revisited. Artif. Intell. **269**, 76–136 (2019)
20. Nigam, V., Jia, L., Thau Loo, B., Scedrov, A.: Maintaining distributed logic programs incrementally. In: PPDP, pp. 125–136 (2011)
21. Qayyum, O., Yu, W.: Toward replicated and asynchronous data streams for edge-cloud applications. In: 37th ACM/SIGAPP Symposium on Applied Computing (SAC), pp. 339–346. ACM (2022)
22. Shapiro, M., Preguiça, N., Baquero, C., Zawirski, M.: Conflict-free replicated data types. In: Défago, X., Petit, F., Villain, V. (eds.) SSS 2011. LNCS, vol. 6976, pp. 386–400. Springer, Heidelberg (2011). https://doi.org/10.1007/978-3-642-24550-3_29
23. Ullman, J.D., Gelder, A.V.: Parallel complexity of logical query programs. Algorithmica **3**, 5–42 (1988)
24. Vogels, W.: Eventually consistent. CACM **52**(1), 40–44 (2009)
25. Xin, R.S., Rosen, J., Zaharia, M., Franklin, M.J., Shenker, S., Stoica, I.: Shark: SQL and rich analytics at scale. In: SIGMOD, pp. 13–24 (2013)
26. Yu, W., Ignat, C.L.: Conflict-free replicated relations for multi-synchronous database management at edge. In: IEEE International Conference on Smart Data Services, pp. 113–121 (2020)
27. Zhang, W., Wang, K., Chau, S.: Data partition and parallel evaluation of datalog programs. IEEE Trans. Knowl. Data Eng. **7**(1), 163–176 (1995)

On-The-Fly Data Distribution to Accelerate Query Processing in Heterogeneous Memory Systems

André Berthold[1(✉)], Lennart Schmidt[2], Antonia Obersteiner[1], Dirk Habich[2], Wolfgang Lehner[2], and Horst Schirmeier[1]

[1] TU Dresden, Operating Systems Group, Dresden, Germany
{andre.berthold,horst.schirmeier}@tu-dresden.de,
antonia.obersteiner@tu-dresden.de

[2] TU Dresden, Database Research Group, Dresden, Germany
{lennart.schmidt1,dirk.habich,wolfgang.lehner}@tu-dresden.de

Abstract. To meet response time and throughput demands, data processing architects continuously adapt query processing systems to novel hardware features. For instance, data processing systems already shifted from disk-oriented to main memory-oriented architectures to efficiently exploit the ever-increasing capacities of main memory. A prominent example for such new developments are emerging memory technologies such as very large caches, high-bandwidth memory (HBM), non-uniform memory access (NUMA) or remote-memory designs like CXL. These memories complement regular DRAM by trading off between properties such as capacity, read/write throughput or access latency. However, these degrees of freedom and their inherent complexity make it difficult for database systems to profit from the new hardware.

Taking HBM – as integrated in the 2023 Intel "Sapphire Rapids" Xeon Max processors – as a most recent example for emerging memory technologies, we present results from microbenchmarks that demonstrate different worker-thread saturation patterns for DRAM and HBM in this paper. Then, we subsequently derive characteristics and a cost model to make dynamic on-the-fly data-distribution/movement decisions. Based on this model, we show that for data processing queries with a specific data-reuse pattern, dynamic data (re-)distribution from DRAM to HBM can decrease end-to-end query run times.

Keywords: Query Processing · Heterogeneous Memory

1 Introduction

The key objective of database systems is to reliably manage data, where high query throughput and low query latency are core requirements [1]. To satisfy these requirements, database systems constantly adapt to novel hardware features [3–5, 7, 13, 16, 19]. In the last decades, we have seen numerous advances, in

J. Tekli et al. (Eds.): ADBIS 2024, LNCS 14918, pp. 170–183, 2024.
https://doi.org/10.1007/978-3-031-70626-4_12

particular with respect to memory, processing elements, and interconnects. Currently, the area of memory and storage has seen an unprecedented change: Novel and already commercially available technologies have blurred the mental picture of a memory/storage hierarchy. For example, Non-Volatile RAM (NVRAM) is a prominent representative to question the long-standing memory hierarchy reflected in all system-level applications [17]. However, very large caches, High-Bandwidth-Memory (HBM) [12], Non-Uniform Memory Access (NUMA), or even remote-memory designs (like RDMA or CXL [21]) as well as extremely fast SSDs [10] add to the heterogeneous portfolio of available memory/storage technologies positioned in a multi-dimensional design space. Unfortunately, this degree of freedom makes it difficult for database systems to efficiently use these heterogeneous memories to satisfy the aforementioned core requirements.

To tackle this challenge, a possible general solution is to *statically* partition the data across the heterogeneous memories as e.g., done in NUMA systems [14,18,20]. Then, the partitions are exclusively accessed by an assigned worker (e.g., a thread pinned to a specific hardware thread) for an efficient query processing. This approach is characterized by the fact that data is not moved in the system, but instead the *operations* move to the data partitions. The efficiency of this approach of course depends on the chosen partitioning scheme. To determine a proper data partitioning scheme, technological properties such as memory bandwidth, latency, or capacity as well as data access frequencies (e.g., hot or cold data) must be taken into account. However, since access frequencies may change over time, this static partitioning approach also has considerable disadvantages.

Core Contribution: To overcome these disadvantages, we envision an alternative general solution called *dynamic On-the-fly Data Distribution (ODD)*. Our overall vision is inspired by the fact that heterogeneous memories generally differ in terms of memory bandwidth, latency, and capacity. For example, Intel's newest Xeon Max CPUs (Sapphire Rapids architecture) feature regular DRAM and HBM, whereas HBM is limited to 64 GiB per CPU but offers a much higher bandwidth – and interestingly also higher access latencies – compared to regular DRAM. Based on these differences, our alternative solution ODD considers HBM as a *memory accelerator* for DRAM-oriented query processing. We assume that all base data (columns) are stored in DRAM, and that during query processing, parts of these data – which are accessed later in the query execution plan – are explicitly on-the-fly distributed to HBM to accelerate subsequent query operators as well as to speed up the overall end-to-end query processing.

Detailed Contributions and Outline: In this paper, we present a first but concrete initial step for our envisioned and novel *ODD* approach. In particular, we make the following detailed contributions:

1. In Sect. 2, we present microbenchmark results of a recent "Sapphire Rapids" Xeon Max processor. In particular, we show that this processor has different worker-thread saturation patterns for DRAM and HBM, but that there is no significant difference for copying data between both memories when not exceeding five worker threads.

2. Based on these microbenchmark results, we subsequently derive characteristics and an initial cost model to make dynamic on-the-fly data distribution (ODD) decisions in Sect. 3.
3. To verify this initial cost model and to evaluate our ODD approach, we describe a prototypical implementation of two representative queries. Then, we empirically show the potential of ODD by presenting evaluation results for both queries with varying the base-data size in Sect. 4.

We conclude with related work and a brief summary including future work in Sects. 5 and 6.

2 Microbenchmarks

In this section, we introduce our underlying hardware setup and present memory throughput results including corresponding lessons learned.

Hardware Setup: For this paper, the fundamental hardware setup is a new two-socket system with Intel Xeon CPU Max 9468 processors (Sapphire Rapids architecture). These CPUs are equipped with 48 cores per socket and two hyper-threads per core [9]. Furthermore, the system offers two integral types of main memory, HBM and DDR5 DRAM adding up to 64 GiB and 256 GiB per socket, respectively. Our first measurements showed that the access latency of 156.33 ns for HBM is higher than the access latency of 132.57 ns for DRAM.

Moreover, we configured the system to use *1-Level Memory* (1LM) mode, meaning that both HBM and DRAM are exposed to the system as separate NUMA nodes and explicitly accessible. Additionally, we activated the *Sub-NUMA clustering-4* (SNC4) mode on both sockets splitting each processor into four logical NUMA nodes, dividing the cores and the DRAM equally among them. Moreover, HBM is divided as well and exposed as four additional NUMA nodes. The SNC4 cluster mode should lead to higher throughput and lower latency compared to *Quadrant* mode, where the processor is not subdivided [8]. To avoid effects that attribute to NUMA clustering, we decided to carry out all experiments on only one NUMA node, which leaves us with twelve cores, 64 GiB of locally-attached DRAM, and 16 GiB of locally-attached HBM. Additionally, all experiments are implemented using AVX-512 instructions to maximize the data throughput of every core, aiming to saturate the system's memory throughput with as few cores a possible. To prevent hyper-threading and thread migration, we pin every thread to its own core by setting the CPU affinity with `cpu_set`.

Memory Read Throughput: Figure 1(a) depicts the achieved results of our benchmark measuring the read throughputs of the system's memories. In the benchmark, we measured the throughput of reading exclusively from DRAM, exclusively from HBM, and reading from both memories in parallel (Combined) over a period of two seconds. As one thread cannot saturate the memories' throughput alone, we tested how the throughput develops with growing thread counts. To see how hyper-threading affects the throughput, we also measured thread counts between 13 and 24.

When examining the **DRAM throughput**, we observe linear growth up to three threads before the increase flattens, maxes out at seven threads, and then even slightly decreases. As running more than five threads reading DRAM brings no notable throughput gains, we take the DRAM's throughput as saturated with five threads. In contrast to DRAM, the **HBM's throughput** shows a nearly linear increase until nine threads are reading concurrently, and reaches a plateau with twelve threads. Increasing the thread count further brings no serious throughput gains, so to saturate the HBM's throughput it is not necessary to use hyper-threads. On the contrary, when scheduling too many threads, the throughput drops by nearly 15%. It is particularly interesting that the maximum HBM throughput is more than twice as high as the DRAM's maximum.

To investigate concurrent read accesses to DRAM and HBM, the **combined throughput** starts with five threads reading from DRAM as the most beneficial configuration for DRAM. The subsequent threads only read from HBM. As the thread count increases, we first see a linear growth of the throughput in Fig. 1(a). Increasing the thread count further, the throughput approaches the measured HBM throughput but does not exceed it, regardless of the thread count. We performed the same measurements with different numbers of threads reading from DRAM, all with a similar observation that we cannot surpass HBM when reading from both memory types at once.

Consequently, for optimal throughput it should be sufficient to accommodate the database in HBM, which is made more difficult by the limited size of the HBM. If the database is too large, placing *parts* of it in HBM could already be beneficial, but we use large amounts of the memory to store columns (or fractions of them) that are not accessed by a query. Thus, we are looking for a *dynamic* approach that on-the-fly copies only the accessed data to HBM during query processing. To get an idea about the involved cost, we examine the copy throughput next.

Copy Throughput: To benchmark the system's copy throughput, we copy 15 GiB of data[1] from DRAM to HBM and vice versa. The data-copy implementation loads the data from one memory into an AVX-512 register and then stores it to the other memory. Loading and storing of data can be done in different ways. We implemented the algorithm (i) with streaming instructions (Stream), which avoid the cache structure, and (ii) with aligned load and store (Aligned), which do not avoid the cache. Figure 1b shows the copy throughput with an increasing number of copy threads. When copying data from **DRAM to HBM**, the different load and store instructions barely make a difference. The maximum copy throughput is 63 GiB/s, which matches the DRAM read throughput. However, with the same number of threads we cannot achieve the same throughput values as in the read throughput benchmark (Fig. 1a). For copying data from **HBM to DRAM**, Fig. 1b shows two very different graphs. The variant using aligned load

[1] slightly below the available HBM of 16 GiB per NUMA node to avoid an out-of-memory condition of unknown origin we experience on Linux.

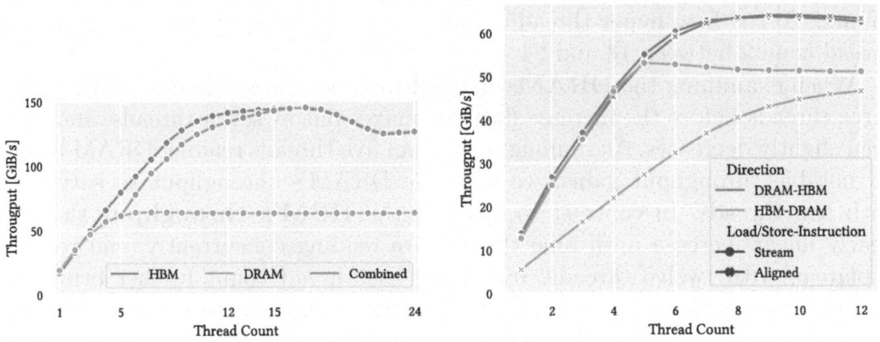

(a) Read throughput.

(b) Copy throughput.

Fig. 1. Microbenchmark results.

and store shows a similar but much flatter throughput curve, so that the peak is still not reached with twelve cores. When using streaming instructions, we reach throughput values similar to the ones achieved when copying from DRAM to HBM. However, when exceeding five threads, the climb comes to a sudden end, and the throughput slightly decreases with higher thread counts. We attribute this to the DRAM's write throughput, which appears to be around 53 GiB/s.

Apparently, the direction in which we copy makes no difference when not exceeding five threads. However, it should be considered using streaming instructions when copying from HBM. To summarize, if HBM is available, throughput-wise it is definitely worth using it as the only database-column storage. If the HBM size is a limiting factor, distributing the data between HBM and DRAM still delivers a higher throughput than processing exclusively in DRAM. When applying a dynamic approach that copies and processes data concurrently – like envisioned with ODD (see Sect. 3), – an important lesson learned is that copying threads take away less memory throughput than reading threads.

3 On-The-Fly Data Distribution

This section introduces *On-the-fly Data Distribution* (ODD), an initial approach to explicitly copy data during query processing while not increasing the query execution time. First, we describe two example scenarios to illustrate our approach (Sect. 3.1). Then, we explain ODD on the basis of these examples (Sect. 3.2). Finally, we introduce a model that approximates the execution times of the example queries under given system properties (Sect. 3.3).

3.1 Scenario

To explain and also to evaluate our envisioned ODD approach, we introduce two fairly simple queries as shown in Fig. 2. Both Query 1 and 2 work on a

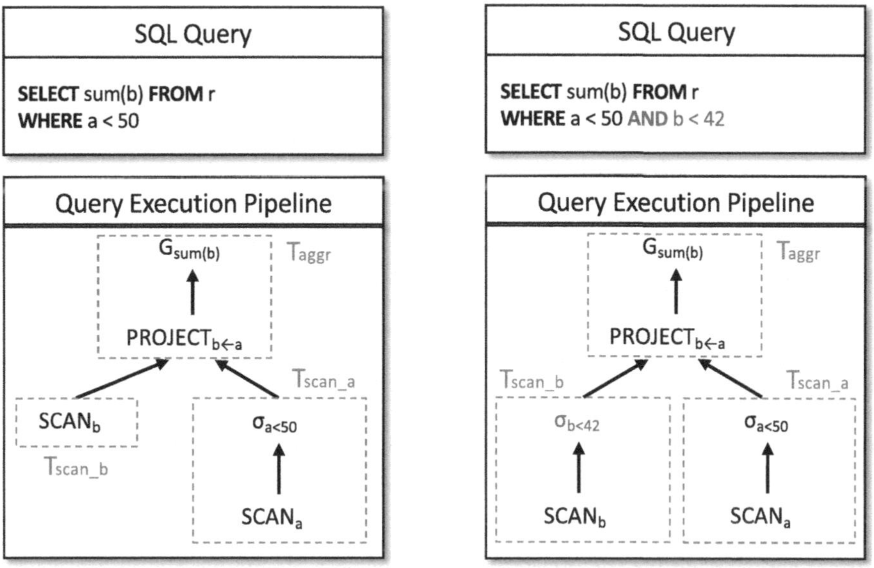

(a) Query 1. (b) Query 2.

Fig. 2. Two example queries with task boundaries that divide the pipeline in three dependent parallelizable tasks (differences left/right highlighted). (Color figure online)

database table that contains two columns, r.a and r.b. ODD is of course not limited to these queries, but they form comprehensible scenarios whose execution times are demonstrably not extended by ODD. Query 2 (Fig. 2b) extends Query 1 with a further filter operator. We assume these queries to be executed in an operator-at-a-time manner [4], so both queries are subdivided into three tasks, T_{scan_a}, T_{scan_b}, and T_{aggr}, which are executed sequentially. Furthermore, we assume that both, Column r.a and Column r.b, according to the in-memory database concept, already reside in DRAM when the query arrives, so the operators $SCAN_a$ and $SCAN_b$ are no-ops. Importantly, the depicted PROJECT operator transfers the result of a selection on one column to another column. In both queries, a SUM-aggregation is carried out after this PROJECT operator.

3.2 Approach

The ODD approach is based on the presented observation in Sect. 2 that a single DRAM device can only sufficiently supply a limited number of cores with data. As shown in Fig. 1a, we achieve best throughput by reading data from HBM alone or from both memories simultaneously. Thus, we must copy at least a subset of the data to HBM. To avoid increasing the query execution time by the copy overhead, we could statically place data in HBM, but as we have no information about the query, we probably waste huge amounts of HBM to unused

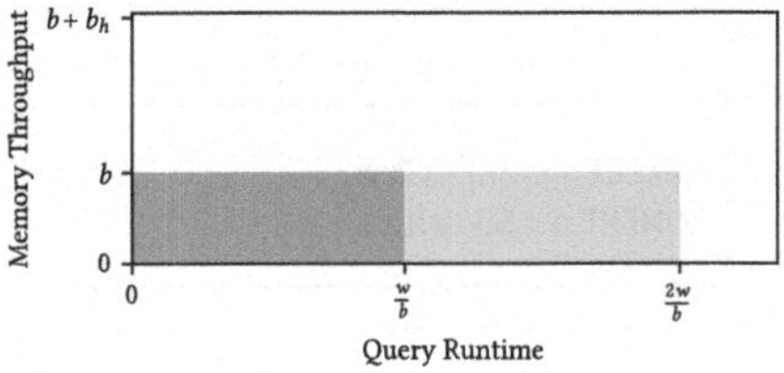

(a) Execution of Query 1 without data distribution.

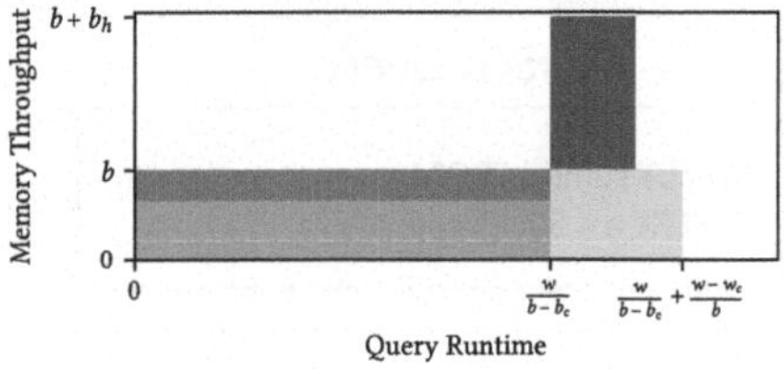

(b) Execution of Query 1 using ODD to distribute Column r.b.

Fig. 3. Illustration of the execution of Query 1 without and with ODD. Every horizontal bar represents a thread, tasked with T_{scan_a} (light blue), T_{aggr} on data in DRAM (light green), data distribution (dark blue), or T_{aggr} on data in HBM (dark green). The height of the bar visualizes the thread's throughput and the width its execution time. It can be seen that execution takes the same amount of time in both cases. (Color figure online)

data. When waiting for the query to arrive, the cost of copying the columns used by the query to HBM is part of the query execution time and the data size could collide with HBM-size limitations.

ODD does not aim at copying *all* data for the arriving query but only a fraction, and then read HBM and DRAM simultaneously to achieve a higher throughput. During an early query-processing phase, two types of threads are active: One makes actual progress, and the other distributes data that are used in a later query-processing phase. While copying, the two types of threads compete for the limited DRAM throughput, which leads to a longer processing time compared to processing the data normally. In the later processing phase, we can profit from the data distribution, as we can schedule additional threads that work on the copied data in HBM and are not bound to the DRAM's read throughput. Ideally, this at least amortizes the increased execution time of the early, copy-slackened phase.

For a better understanding, Fig. 3 visualizes the execution of Query 1 without and with data distribution. In Fig. 3a, we can only harness the read throughput b provided by DRAM, for both, T_{scan_a} and T_{aggr}. During execution, we read Column r.a and Column r.b, which both have a size of w in our scenario, so have an overall execution time of $\frac{2w}{b}$. When applying ODD (Fig. 3b), we reduce the throughput of T_{scan_a} to schedule threads that concurrently read Column r.b from the same DRAM device and store it in HBM. Consequently, T_{scan_a} can only read the workload w with the DRAM throughput b minus the throughput of the copying threads b_c, resulting in a execution time of $\frac{w}{b-b_c}$. Of course, we need to ensure that $b_c \leq \frac{b}{2}$, otherwise the copying threads would take up too much of the memory throughput, such that processing of T_{scan_a} alone would take longer than the normal processing of Query 1. The copy throughput can be regulated by the number of copying threads. We scale the amount of data to be copied, such that the copying threads' execution time matches with the execution time of $T_{\text{scan\_a}}$, allowing us to copy $w_c = \frac{w}{b-b_c} b_c$ of Column r.b. As Column r.b is distributed, we can schedule more threads for T_{aggr} utilizing the DRAM's and HBM's throughputs $(b+b_h)$. Thus, the workload for the in-DRAM processing of T_{aggr} is reduced by w_c. So, T_{aggr} takes either the time to process the reduced workload in DRAM or the time to process the copied workload in HBM $(\max(\frac{w-w_c}{b}, \frac{w_c}{b_h}))$.

In the following equations, we want to show that the execution of Query 1 with (right hand side of the equations) and without ODD (left hand side) takes the same time, assuming that $\frac{w-w_c}{b} \geq \frac{w_c}{b_h}$ (scenario depicted in Fig. 3b):

$$\begin{aligned}
\frac{2w}{b} &= \frac{w}{b-b_c} + \frac{w-w_c}{b} && \Big| w_c \hat{=} \frac{w}{b-b_c} b_c \\
\frac{2w}{b} &= \frac{w}{b-b_c} + \frac{w - \frac{w}{b-b_c} b_c}{b} && \Big| \cdot b \\
2w &= \frac{wb}{b-b_c} + w - \frac{wb_c}{b-b_c} && \Big| - w \\
w &= \frac{wb - wb_c}{b-b_c} && \Big| \cdot (b-b_c) \\
wb - wb_c &= wb - wb_c
\end{aligned}$$

As it turns out this assumption holds if we ensure that $b_c \leq \frac{b \cdot b_h}{b+2b_h}$ is satisfied. The equations show that the overall execution time of the Query 1 does not increase, compared to execution without ODD. However, without losing execution time, we change the system state in the process, such that a part of Column r.b is now stored in HBM: Further accesses to this column will benefit from the increased throughput. Consequently, applying ODD to Query 2 should reduce overall execution time. The copy penalties for $T_{\text{scan\_a}}$ are already amortized during execution of $T_{\text{scan\_b}}$, such that T_{aggr}'s reduction in execution time directly contributes to the overall execution time reduction.

3.3 Model

We used the gathered data from the microbenchmarks (Sect. 2) to develop a model that helps to approximate the optimal number of copy threads and the best value for w_c for our benchmark (Sect. 4).

The main query operators, performed in the introduced Queries 1 and 2, are summation and filter operations. These have low computational effort, and we assume their processing time is bound to the read throughput of the corresponding memory device. Furthermore, in our model we neglect the computational effort and memory pressure of the $\text{PROJECT}_{\text{a} \leftarrow \text{b}}$ operator used in the queries. In the following calculation, parameters n_d, n_h, and n_c determine the number of threads working on data in DRAM/HBM, and distributing data, respectively. Hereby, n_c is not necessarily integral and is included in n_d, so the higher n_c the fewer threads work on the completion of $T_{\text{scan\_a}}$. Given these assumptions and parameters, we can approximate the execution times of Query 1 and 2 without ODD as follows:

$$t_{Q1}(n_d) = \frac{2w}{b(n_d)} \qquad t_{Q2}(n_d) = \frac{3w}{b(n_d)}$$

We can also calculate the approximations *with* ODD:

$$\begin{aligned}
w_c(n_d, n_c) &= \frac{w}{b(n_d) - b_c(n_d, n_c)} b_c(n_d, n_c) \\
t_{Q1,ODD}(n_d, n_c, n_h) &= \frac{w}{b(n_d) - b_c(n_d, n_c)} + \\
&\max\left(\frac{w - w_c(n_d, n_c)}{b(n_d)}, \frac{w_c(n_d, n_c)}{b_h(n_h)}\right) \\
t_{Q2,ODD}(n_d, n_c, n_h) &= \frac{w}{b(n_d) - b_c(n_d, n_c)} + \\
&2 \cdot \max\left(\frac{w - w_c(n_d, n_c)}{b(n_d)}, \frac{w_c(n_d, n_c)}{b_h(n_h)}\right)
\end{aligned}$$

Functions $b : \mathbb{N} \to \mathbb{R}$ and $b_h : \mathbb{N} \to \mathbb{R}$ define the read throughput when accessing DRAM or HBM, respectively, with the specified number of threads. We can extract the necessary information for both functions from the benchmarks presented in Sect. 2. Function $b_c(n_d, n_c) : \mathbb{N} \times \mathbb{R} \to \mathbb{R}$ defines the throughput when copying with n_c threads, assuming in total n_d threads are accessing the source memory device. For this function we have no dedicated microbenchmark, thus, we used the results of the microbenchmark measuring copy throughput. To approximate b_c we multiplied the copy throughput measured for n_c by $\frac{n_c}{n_d}$.

In Fig. 4, we see the model's estimates for the execution times of the two example queries. According to our test setup and observations from Sect. 2, we set the parameters $n_d = 5$ and $n_h = 12 - n_d = 7$, and dimension the workload with 15 GiB. The orange line shows the execution time of the corresponding query without ODD, while the blue line approximates execution time with a growing number of copy threads.

In Fig. 4a, we can see the approximations for Query 1. In the beginning, we observe a constant execution time. In this range the inequality $b_c(n_d, n_c) \leq \frac{b(n_d) b_h(n_h)}{b(n_d) + 2 b_h(n_h)}$ holds. For higher numbers of copy threads either the amount of copied data is so high that execution in HBM bottlenecks $\mathrm{T_{aggr}}$, or the memory throughput used for $\mathrm{T_{scan\_a}}$ gets too small. In contrast, Fig. 4b shows the promising results for Query 2. As the number of copying threads increases, we can initially achieve lower execution times, because later tasks also benefit from the copied data in HBM. But when exceeding two copy threads, the execution time increases due to the negative effects explained above for Query 1.

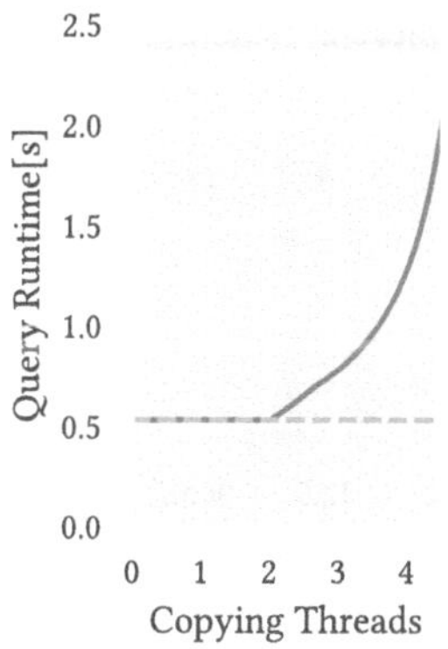

(a) Execution time of Query 1 using ODD, showing the same execution time as the baseline for two or less copying threads.

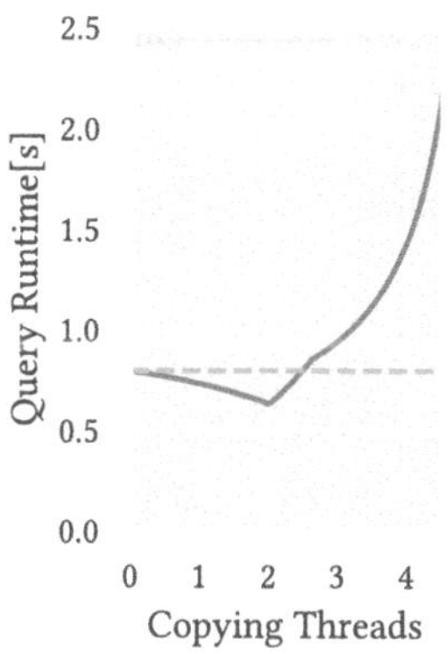

(b) Execution time of Query 2 using ODD, outperforming the baseline, showing the best execution time with two copying threads.

Fig. 4. The model's execution time estimates for both queries using ODD. The baseline (orange) is their execution time without ODD. The number of copying threads is varied, which alters the copy throughput b_c. (Color figure online)

4 Evaluation

In this section we check if the theoretically discussed properties from Sect. 3 still hold for real hardware. First, we describe important aspects of our implementation of ODD. Then, we present benchmarking results and discuss them.

4.1 Implementation

In Sect. 2, we found that hyper-threads do not increase the achievable system throughput, so we only schedule one thread per core (max. 12 at a time). For calibration, we use the model described in Sect. 3.3, which calculates the optimal number of copying threads and the fraction of data that should be copied to HBM. For compilation, we used GNU g++ version 13.2.0 in its C++20 mode. To ensure optimal performance and system compatibility, we applied the compiler flags `-O3` and `-march=sapphirerapids`.

As high-level language abstraction for threads, we used the standard library implementation of C++, `std::thread`. We started the threads synchronously using `std::future`. Before the threads responsible for the following task were started, all threads of the previous task were joined (`std::thread::join`) into the main thread. We decided to load and store the data to and from the AVX-512 registers using streaming instructions to keep the measurements consistent with our microbenchmark measurements. For memory management we primarily used `numa_alloc_onnode` and `numa_free` from `libnuma`.

The workload consists of uniformly distributed 64-bit integers between 0 and 100. For time measurements, we used the `std::chrono::steady_clock`

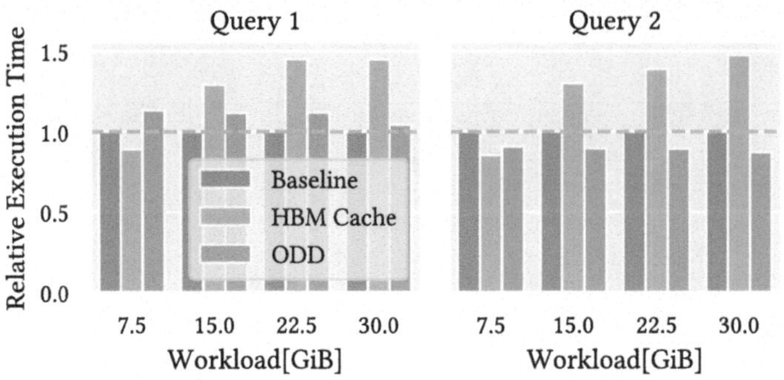

Fig. 5. Execution time of the Queries 1 and 2 executed under 2LM, Baseline conditions and with ODD. 2LM is best, when both columns r.a and r.b fit into HBM. For higher workloads, the baseline performs best for Query 1, and ODD performs best for Query 2.

from the C++ standard template library. We set up the measurements in a way to only measure the processing of the workload. So memory management, data initialization, thread creation, etc. happens outside of the measurement's scope. The measurement starts directly before starting execution of the first set of threads (responsible for the first task) and ends right after we retrieve the query's result. To increase measurement accuracy, we ran the benchmark on an unloaded system, every benchmark parameter configuration was executed 30 times, and we set Linux 6.6.13-amd64's CPU governor to `performance`.

For comparability we benchmarked three different modes, *Baseline*, *On-the-fly Data Distribution* (ODD), and *HBM Cache mode*. Baseline is the execution of the query exclusively in DRAM, as illustrated in Fig. 3a. In ODD mode, we applied our ODD to the query, as described in Secion 3.2. The HBM Cache mode is the exception to the aforementioned setup descriptions. For this measurement, we switched to the *2-level memory* (2LM) mode in the BIOS settings and carried out the Baseline measurements with this hardware configuration. Furthermore, we needed to use aligned load and store intrinsics, in order to not bypass the caches but show their effect. The 2LM mode was introduced by Intel, so that old, HBM-unaware code can benefit from its special properties [8].

4.2 Benchmark Results

Query 1: Figure 5 shows the results of our benchmark implementation. We performed both queries with different workload sizes per column (x-axis) and in the aforementioned three configurations (modes). On the y-axis, we can see the execution time of the queries relative to the workload's Baseline mode.

In the left diagram, the benchmark results of Query 1 are depicted. When comparing the HBM Cache mode to the Baseline, we observe that for the smallest workload size (7.5 GiB) – where both columns a and b fit into HBM together, – HBM Cache achieves better runtimes. But as the workload grows the additional costs of moving **all** accessed data to HBM cannot be amortized by the higher throughput. This also explains the difference between the ODD approach, which knowingly avoids to copy all data to HBM, and HBM Cache.

But as already discussed in Sect. 3.2, for Query 1 ODD cannot outperform the Baseline. We showed that theoretically, Baseline and ODD would have the same execution time. We attribute the fact that our measurements (Fig. 5) prove this wrong to inaccuracies within our model, which was used to define the parameters – specifically the amount of data to copy and the number of copy threads. Furthermore, our benchmark code allows no fractions of copy threads.

Query 2: On the right hand side of Fig. 5, we see the results for Query 2. For HBM Cache, we see a similar picture as for Query 1. Comparing ODD with the Baseline, we see that it constantly outperforms the Baseline by around 10%. And even for the smaller workloads, when HBM Cache performs best, we can reach a comparably good runtime.

4.3 Discussion

For Query 1, we have seen an increase in execution time compared to the baseline, which contradicts our model's approximations and the proof from Sect. 3.2. We explain this by the simplicity of our initial model. The assumption that the filter and sum operations are equally complex in memory pressure might be too negligent, as well as neglecting the computational effort of the operators. Additionally, we ignored the $\mathrm{PROJECT}_{a \leftarrow b}$ operator in the model completely. This also explains the differences between the approximated (20%) and actual (10%) execution time improvements for Query 2. An imprecise model not only returns wrong approximations, but also falsely parametrizes the benchmark, increasing the differences of estimates and measurements.

The execution time improvement of 10% might seem insignificant at first glance. However, Query 2 actually represents a fairly challenging scenario. During the first query processing phase $\mathrm{T}_{\mathrm{scan\_a}}$, a very data hungry task that barely does any computations, competes with the copy tasks for throughput. If during the on-the-fly data distribution, a task with more computational efforts was executed, the unused throughput could have been used for even more copying.

5 Related Work

Generally, developments in the hardware sector have a major impact on database systems. This applies to processing units (CPUs [18,23], GPUs [5], FPGAs [11], ASICs [2]), memory [3,7,16,17,24], and interconnects [6,15,22,24]. In the area of memory, for example, non-volatile memory (NVM) revolutionized the architecture of database systems [17]. NVM is byte-addressable and exhibits latencies similar to those of DRAM, while being non-volatile. Moreover, some recent work started to analyze the impact of High-Bandwidth-Memory (HBM) on join processing [19]. In line with this research direction, we present a novel approach by considering HBM as a *memory accelerator* for DRAM-oriented query processing.

6 Conclusion

In this paper, we carefully examined the achievable memory throughput within one NUMA node of a hybrid DRAM/HBM system. We found that even though accessing DRAM and HBM in parallel, the throughput of accessing HBM exclusively cannot be exceeded. We introduced a novel data distribution approach, ODD, to accelerate database query processing, which on-the-fly copies data to HBM, allowing higher memory throughput during a later processing phase. Based on microbenchmarking results, we developed a model to approximate the query-execution time and parametrize query processing. Lastly, we benchmarked the ODD approach, showing that even for a challenging scenario, execution-time improvements of 10% are possible.

Based on the demonstrated benefits of ODD, future work is manifold. On the one hand, the cost model must be refined in order to calibrate the ODD approach more precisely. It should be adapted to the use case of concurrent query processing, which we intend to investigate in further research. On the other hand, more complex queries must be considered. In addition, it should be investigated whether the ODD approach can be generalized to cover several heterogeneous memory environments.

Acknowledgements. We gratefully acknowledge Dr. Alexander Krause for an initial microbenchmark setup to perform accurate measurements. This work was partly supported by the German Research Foundation (DFG) priority program SPP 2377 under grants no. SCHI 1447/1-1 and LE 1416/30-1.

Disclosure of Interests. The authors declare no competing interests.

References

1. Abadi, D., et al.: The Beckman report on database research. Commun. ACM **59**(2), 92–99 (2016)
2. Arnold, O., Haas, S., Fettweis, G.P., Schlegel, B., Kissinger, T., Lehner, W.: An application-specific instruction set for accelerating set-oriented database primitives. In: SIGMOD, pp. 767–778 (2014)
3. Baumstark, A., Jibril, M.A., Sattler, K.: Processing-in-memory for databases: query processing and data transfer. In: DaMoN@SIGMOD, pp. 107–111 (2023)
4. Boncz, P.A., Kersten, M.L., Manegold, S.: Breaking the memory wall in MonetDB. Commun. ACM **51**(12), 77–85 (2008)
5. Breß, S., Funke, H., Teubner, J.: Robust query processing in co-processor-accelerated databases. In: SIGMOD, pp. 1891–1906 (2016)
6. Geyer, A., et al.: Near to far: an evaluation of disaggregated memory for in-memory data processing. In: DIMES@SOSP, pp. 16–22 (2023)
7. Hedam, N., Clausen, M.T., Bonnet, P., Lee, S., Larsen, K.F.: Delilah: eBPF-offload on computational storage. In: DaMoN@SIGMOD, pp. 70–76 (2023)
8. Intel: Intel Xeon CPU Max Series – Configuration and Tuning Guide. Intel (2023)
9. Intel: Intel Xeon CPU MAX 9468 processor (2024). https://www.intel.com/content/www/us/en/products/sku/232596/intel-xeon-cpu-max-9468-processor-105m-cache-2-10-ghz/specifications.html

10. Jeong, J., Park, K., Lee, S., Bonnet, P., Lerner, A., Cudré-Mauroux, P.: Far-and-near: Co-designed storage reliability between database and SSDs. In: CIDR (2023)
11. Jiang, W., Korolija, D., Alonso, G.: Data processing with FPGAs on modern architectures. In: SIGMOD, pp. 77–82 (2023)
12. Jun, H., et al.: HBM (high bandwidth memory) DRAM technology and architecture. In: 2017 IEEE International Memory Workshop (IMW), pp. 1–4 (2017)
13. Karnagel, T., Habich, D., Lehner, W.: Adaptive work placement for query processing on heterogeneous computing resources. Proc. VLDB Endow. **10**(7), 733–744 (2017)
14. Kissinger, T., Kiefer, T., Schlegel, B., Habich, D., Molka, D., Lehner, W.: ERIS: a NUMA-aware in-memory storage engine for analytical workload. In: ADMS@VLDB, pp. 74–85 (2014)
15. Lerner, A., et al.: Databases on modern networks: a decade of research that now comes into practice. Proc. VLDB Endow. **16**(12), 3894–3897 (2023)
16. Li, F., Das, S., Syamala, M., Narasayya, V.R.: Accelerating relational databases by leveraging remote memory and RDMA. In: SIGMOD, pp. 355–370. ACM (2016)
17. Oukid, I., Booss, D., Lespinasse, A., Lehner, W., Willhalm, T., Gomes, G.: Memory management techniques for large-scale persistent-main-memory systems. Proc. VLDB Endow. **10**(11), 1166–1177 (2017)
18. Pandis, I., Johnson, R., Hardavellas, N., Ailamaki, A.: Data-oriented transaction execution. PVLDB **3**(1), 928–939 (2010)
19. Pohl, C., Sattler, K., Graefe, G.: Joins on high-bandwidth memory: a new level in the memory hierarchy. VLDB J. **29**(2–3), 797–817 (2020)
20. Psaroudakis, I., Scheuer, T., May, N., Sellami, A., Ailamaki, A.: Adaptive NUMA-aware data placement and task scheduling for analytical workloads in main-memory column-stores. PVLDB **10**(2), 37–48 (2016)
21. The CXL Consortium: Compute express link. https://www.computeexpresslink.org/ (2019). Accessed 22 Mar 2023
22. Thostrup, L., Doci, G., Boeschen, N., Luthra, M., Binnig, C.: Distributed GPU joins on fast RDMA-capable networks. Proc. ACM Manage. Data **1**(1), 29:1–29:26 (2023)
23. Ungethüm, A., et al.: Hardware-oblivious SIMD parallelism for in-memory column-stores. In: CIDR (2020)
24. Ziegler, T., Binnig, C., Leis, V.: ScaleStore: a fast and cost-efficient storage engine using DRAM, NVMe, and RDMA. In: SIGMOD, pp. 685–699 (2022)

Enhancing Machine Learning Capabilities in Data Lakes with AutoML and LLMs

Sayed Hoseini[1(✉)], Maximilian Ibbels[1], and Christoph Quix[1,2]

[1] Niederrhein University of Applied Sciences, Krefeld, Germany
sayed.hoseini@hs-niederrhein.de
[2] Fraunhofer FIT, St. Augustin, Germany

Abstract. The exponential growth of data from digitization requires efficient utilization and storage of large amounts of data. Data lakes can store heterogeneous datasets and prepare them for machine learning (ML). However, current data lakes lack mature capabilities to support ML requirements. AutoML is the process of automating the end-to-end application of ML to real-world problems. Large Language Models (LLMs) can potentially increase ML pipeline automation by assisting at various stages of the process and democratizing access to advanced analytics. This paper explores the integration of AutoML tools and LLMs and their application in the data lake *SEDAR*. We present an extended data lake metadata model for capturing data analytics, a Python package for wrapping AutoML libraries, and a module that leverages LLMs for AutoML. Finally, we undertake a comparative analysis between the performance of AutoML and LLMs in four challenging real-world use cases from the domain of chemistry, each presenting a distinct type of ML problem.

Keywords: Data Lakes · AutoML · LLMs

1 Introduction

The exponential growth of data resulting from digitization requires efficient utilization of the information obtained. Machine data analysis and pattern recognition find broad applications in nearly all economic and research sectors [15]. For both, the analysis and the management of such vast data assets, data lakes (DL) as schema-agnostic repositories for storing data in their native format are proposed. They provide centralized access and the capability to apply data transformation when needed according to an Extraction-Load-Transformation approach [5]. While data is largely consumed by machine learning (ML) and data science applications everywhere, existing DLs lack matured functions to support this requirement [7]. Data lake metadata often do not include information for ML pipelines particularly [28].

MLOps [1] encompasses many practices for deploying, maintaining, and managing ML models easily, efficiently, and reliably. Within an MLOps framework, data collection, analysis, and cleaning can be organized. Additionally, version

J. Tekli et al. (Eds.): ADBIS 2024, LNCS 14918, pp. 184–198, 2024.
https://doi.org/10.1007/978-3-031-70626-4_13

control for data is provided, the model and underlying program code are stored, the ML model can be automatically deployed or tested, and it can be evaluated whether the model performs well and results are reproducible. ML-driven metadata management includes multiple steps, such as model training, hyperparameter tuning, debugging, deployment, and more. Apart from providing the relevant descriptive metadata of data sources in the lake, coverage of the metadata of ML models based on this data is thus required also [7,20,28].

The multitude of ML techniques, the extensive steps involved in deploying an ML model for a business application, and the need for statistical and technical knowledge all pose significant challenges in deriving value from data for non-technical users. While the demand for analyses is increasing, existing solutions are still often manually created by humans [15]. Automated machine learning (AutoML) deals with the automation of individual steps within the ML process. This automation is ultimately limited by the fact that the work of an analyst, especially in the phase of problem definition and data preparation, involves correspondence with domain experts. However, current AutoML systems are capable of assisting in the substeps of model creation, i.e., model selection and hyperparameter tuning [15].

In wake of the rise of Large Language Models (LLMs), we expect a major impact on the automation of ML pipelines. LLMs, such as *GPT-3.5*, have demonstrated remarkable capabilities in understanding, generating, and processing vast amounts of textual data [17]. LLMs can assist in every stage, e.g., data preprocessing, feature engineering, model selection, hyperparameter optimization, and result interpretation. By leveraging the vast knowledge encoded within LLMs, AutoML systems can accelerate the process of building and deploying ML models, democratizing access to advanced analytical tools. LLMs may serve as assistants that empower users with diverse backgrounds to harness the full potential of ML without extensive technical expertise.

In this paper, we address the question of how the integration of automated machine learning can augment the capabilities of a data lake system and compare the efficacy of AutoML tools with LLMs in this context. Our contributions include:

1. Development of a Python module to integrate various AutoML libraries, designed to handle diverse use cases with a focus on uniform job configuration.
2. Implementation of a specialized module within the *SEDAR* data lake system [10] that leverages LLMs for configuring and automating ML pipelines, enhancing both efficiency and accessibility.
3. Definition of a comprehensive metadata model that captures essential data analytics information throughout different phases of ML within a data lake, facilitating advanced data management.
4. An evaluation of our proposed framework against challenging non-standard real-world tasks within the domain of chemistry, validating its applicability in practical settings.
5. A comparative discussion of the performance of LLM and AutoML techniques.

2 Related Work

A comprehensive survey for data management for ML is presented by Chai et al. [2]. They identify the main tasks for any data management platform to support ML pipelines: data discovery, cleaning, and labeling in the phase of data preparation as well as feature and model selection and hardware acceleration during model training and inference, and finally model storage, versioning, querying, deployment and serving of created models, pipeline debugging during post-ML model management. Hai et al. [7] highlight the importance of supporting ML processes for DLs and describe current challenges. They also underline the importance of ML-driven metadata management. In-lake ML is introduced to support the training and inference process directly inside the DL platform. In today's landscape, this requires integrating new APIs and systems to connect existing DLs with model zoos (repositories of pre-trained models) such as *GPT4ALL* and *HuggingFace*. Finally, there is the data heterogeneity, i.e., not only supporting tabular data, JSON/XML, and graphs, but also other common types such as images, audio, and videos, which is a non-trivial problem, that stretches beyond simply utilizing technologies such as multimedia databases and polystores.

The problem of heterogeneity in training data is also addressed in [6], where the authors study how to integrate disparate collections of datasets for feature augmentation and federated learning. They present *AMALUR*, a unified ML system utilizing the various metadata in databases and DLs (data-related, model & hardware-related) to integrate heterogeneous and distributed data sources to enhance ML pipelines' efficiency. In [20,28] the authors present metadata models for DLs to capture not only descriptive but also analytical information.

The majority of works related to **AutoML LLM** are preprints submitted to *arXiv* or *OpenReview* that are yet to become peer-reviewed, yet we highlight some important findings. A framework that uses *ChatGPT* as a *Virtual Interactive Data Scientist* is presented in [8]. It assists users in conducting AutoML tasks through natural dialogues without requiring in-depth knowledge of the underlying ML processes. They divide the dialogues into different states, data visualization, task formulation, prediction engineering, and result summary and recommendation, and present a set of respective prompts for each state. The research represents a systematic approach to creating a personal data scientist, but the authors admit that there are still severe problems to be addressed, such as erroneous recognition of the correct current state by the LLM.

Several works utilize LLMs for hyper-parameter optimization [18,26,27]. These methods allow an agent to conduct experiments with specific hyper-parameters for ML algorithms to iteratively optimize them based on historical trials while autonomously processing the task information. Chen et al. [3] integrate LLMs as adaptive operators in an evolutionary neural architecture search, Wei et al. [22] deploy LLM-based agents to devise tailored solutions for diverse graph-structured data and learning tasks and Huang et al. [13] introduced a suite of ML tasks for benchmarking AI research agents.

Tornede et al. [21] review the current state of the integration between LLMs and AutoML in both directions. They point out that this integration has the potential to substantially improve the human interaction component of AutoML tools, alleviate the tedious task of correctly configuring the tool which often requires experts, and improve several internal components through knowledge. Currently, interacting with an AutoML system is often challenging for non-expert users. LLMs may fundamentally redesign this interaction, propose an effective initial setup to a particular problem, and even assist in interpreting the output. Simultaneously, the integration also bears risks such as inadequate evaluation of LLM-powered AutoML systems, catastrophically wrong behavior of AutoML systems due to hallucinations of an LLM-powered component, and ever-increasing resource demands. *AutoMMLab* [23] is a LLM-empowered AutoML system that follows the user's language instructions to automate model production workflows for computer vision tasks.

3 Integrating AutoML and LLMs with *SEDAR*

We begin by briefly describing the *SEDAR* data lake system in Sect. 3.1. Then, we present its metadata model which lies at the heart of the system serving as the foundation for all data management functionality in Sect. 3.2. *AutoMLWrapper* is an independent Python module that integrates various AutoML libraries. We present the module in Sect. 3.3. Support for ML tasks by LLMs is implemented as an extension of *SEDAR* (Sect. 3.4) and describes the integration with a special focus on the applicability of AutoML.

3.1 The *Semantic Data Reservoir (SEDAR)*

SEDAR is designed in a modular, easily extendable, and flexible four-layer architecture based on open-source technologies deployed as microservices. To realize a true polystore [16], the *storage layer* is composed of different storage systems to support efficient storage for heterogeneous data structures based on their specific data model. By default, *Hadoop* with its file system *HDFS* is used to store files with arbitrary structures including audio, images, video, text, and ML model artifacts. The *transformation layer* implements various operations for data preparation, e.g., cleansing, transformations, or joins. We employ *Apache Spark* as a unified data transformation engine for large-scale data processing. A programmatic interface to communicate with *SEDAR* is provided by the *SEDAR-API*, an external Python module. To ensure generality, the *ingestion layer* is independent of any formats and data sources, not limited to particular systems in the storage layer, and allows for versioning and the addition of further databases. The *interaction layer* facilitates comprehensive metadata management while ensuring ease of use for non-technical users.

Additionally, modern DLs must aid in creating, managing, and deploying ML applications while also providing necessary data. To facilitate MLOps, *SEDAR* utilizes MLFlow [24], an open-source framework to manage the inputs and outputs of ML applications. This includes tracking model versions, code, and tuning parameters, ensuring reproducibility of results, and facilitating model deployment. *Jupyter Notebooks* have been integrated into the interaction layer and are hosted from a *JupyterHub*, allowing users to have personalized work environments. Authentication between web services is centralized using the OAuth protocol and JWT tokens. Within *SEDAR*, users can select datasets, and then choose the desired learning method, and the system will automatically generate a notebook with the necessary code to load the *DataFrames* from storage via the *SEDAR API*, register their experiments, track individual runs and software environments, store model artifacts, and deploy their applications.

3.2 Metadata Model

A data catalog is fundamental for any organization to manage big data in industrial scenarios. DLs aim to extract as much metadata as possible to populate their data catalog and enable users to enter further details manually. For *SEDAR* we have incorporated several technical requirements like data linkage, indexing, polymorphism, versioning, usage tracking, similarity relationships, and data quality management in addition to the previously described tasks around MLOps and AutoML. This requires comprehensive metadata management.

Figure 1 presents an excerpt of *SEDAR*'s metadata model as the basis for its data catalog to capture relevant information about all processes inside the system to enable search across the various assets. The extension for ML has been inspired by [28]. While the system provides further functionality, like semantic data management [10], this paper focuses on the elements related to analytics. Central to the whole system is the *Dataset* acting as the main entry point. To enable data security and support data governance the concept of *Workspaces* is introduced. Every *User* is assigned to a workspace and may add other users, but generally cannot access (meta-)data from other workspaces unless specific access is granted, or the workspace is made publicly available. Datasets in a workspace can be made public or only to single users. We differentiate between (semi-)structured data (for which the system displays status, extracted schema information, and related datasets), unstructured data as well as images.

To reduce the effort for data preparation, image datasets are subject to certain schema specifications. The *AutoMLWrapper* expects this specification and provides functions for automated pre-processing before feeding it to an individual AutoML framework. The data itself is packaged in a ZIP file and three types of image datasets are considered. Figure 2 contains the schema per task as a file tree. Classification datasets divide the images into subfolders, where the name of the subfolder reflects the class. For object recognition, we use the well-known COCO format, which contains a JSON file with label information. Segmentation problems are supported for foreground-background segmentation, where two subfolders contain the image and the corresponding segmentation mask respectively. Class memberships are generally coded by the pixel values of the masks and images can be uploaded for segmentation with multiple classes. The metadata extraction with *Apache Tika* was configured to obtain specific information about the images of the dataset stored in the *ImageEntity* object

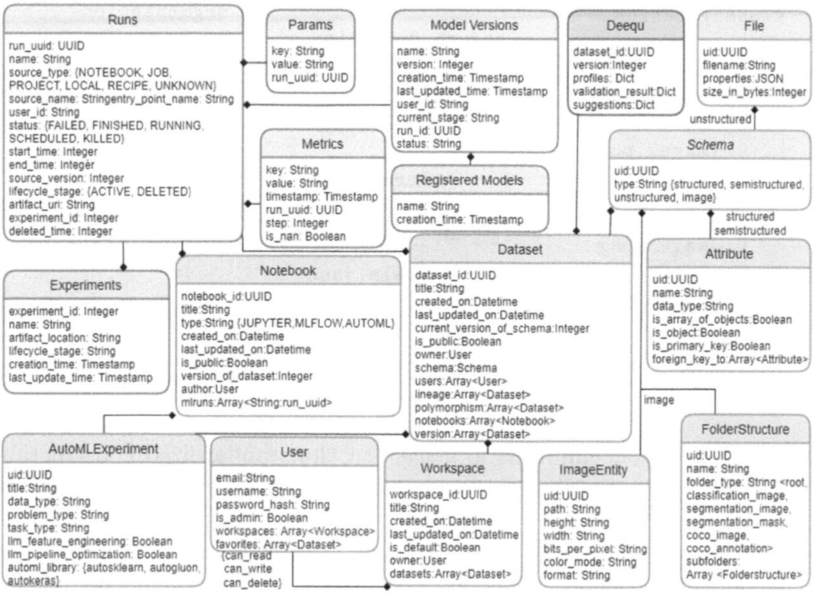

Fig. 1. Excerpt of *SEDAR*'s metadata model to display relevant metadata related to analytics. Green nodes come from *SEDAR* directly, blue nodes are incorporated from *MLFlow*, and red nodes work in connection to *Deequ*.

with basic information about size, format, and data types as well as the *FolderStructure* object to convey labeling information (see Fig. 2).

Because the quality of the training data heavily influences the performance of ML models [14], *SEDAR* utilizes *Amazon*'s *Deequ* [19] library for automating the verification of data quality at scale combining common quality metrics with user-defined validation constraints to enable unit tests for data. The system automatically triggers comprehensive data profiling for any ingested dataset. For reasons of brevity, we have cut this component short and did not include further objects for the profiles dictionary for example.

A dataset can contain one or more ML *Runs* and each belongs to a particular *Experiment*. These runs can either be assigned directly to the dataset via the frontend or there is a *Notebook* between the entities in which a run was registered by code. This run contains the parameters and metrics that the user has individually assigned while creating the model. *MLFlow* further tracks all *Model Versions* and *Registered Models*, i.e. models deployed for production.

3.3 AutoMLWrapper

The *AutoMLWrapper* is a stand-alone Python module acting as an interface for three open-source AutoML frameworks: *AutoKeras*, *AutoGluon* and

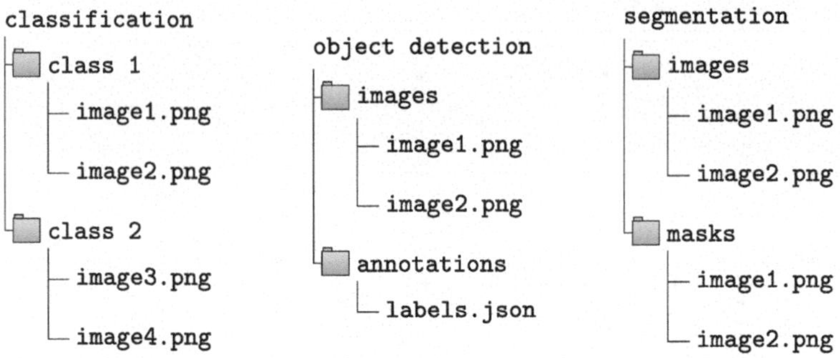

Fig. 2. Schema specification for the ingestion of training images

AutoSklearn[1]. Special emphasis is placed on the standardized construction of AutoML jobs by configuring the underlying libraries using a common set of parameters. Object-oriented programming for library-unspecific logic facilitates easy integration of new AutoML libraries without redesigning the entire interface. To use the package with code as well as in combination with *SEDAR*'s frontend, presenting information about possible settings for AutoML experiments required a form that can be easily understood and displayed. To meet these requirements, each of the three frameworks is equipped with a configuration file essential for the correct call to the underlying library. Then, there is a global configuration file that contains meta-information about the available libraries' scope and assigns data and problem types. For example, the file can be used to decide in the frontend which (library) functions are available to the user

```
AutoGluon:
 __classes:
  MultiModalPredictor:
   __data_types:
    image:
     __task_types:
      classification:
       __problem_types:
        - binary
        - multiclass
      few-shot-classification:
       ...
      zero-shot-classification:
       ...
      zero-shot-object-detection:
       ...
      ...
      segmentation:
       __problem_types:
        - binary
        - multiclass
```

```
AutoGluon:
 MultiModalPredictor:
  __hyperparameter_details:
   constructor:
    ...
    validation_metric:
     __has_values: true
      __values:
       - accuracy
       - f1
       - f1_macro
       - iou
       ...
  fit:
   time_limit:
    __has_values: false
     __dtype: int
     __user_mapping: runtime
     __hint: Time limit for the\
     experiment in seconds.
   ...
```

```
wrapper = AutoMLWrapper('autogluon')
# Enable parameters not present in config files
wrapper.AllowExtraHyperparameters(
    {'constructor': ['verbosity']}
    )

wrapper.Train(
    train_data=train_data,
    validation_data=val_data,
    target_column='label',
    data_type='image',
    task_type='segmentation',
    problem_type='binary',
    hyperparameters={
        'eval_metric':eval_metric,
        'validation_metric':eval_metric,
        'runtime':2700,
        'verbosity':4,
        ...
        }
)
```

Fig. 3. Excerpts of the global (left), the *AutoGluon*-specific configuration file (center) and exemplary code of the *AutoMLWrapper* for image segmentation (right).

[1] https://github.com/automl/auto-sklearn---/keras-team/autokeras---/autogluon/autogluon.

for image segmentation. An excerpt from these files is shown in Fig. 3 alongside a code example.

The *AutoMLWrapper* is integrated into *SEDAR* within a dedicated virtual Python environment built into the *Jupyterhub*'s default image. Similar to the previous manual training scripts a complete *Jupyter Notebook* is created and populated with information from the frontend. By design, the package also integrates with *MLFlow* to manage experiments and save the generated models. Despite the idea of automation, the package also provides enough freedom for experienced users allowing them to add their own logic and specific customizations. This notebook serves as the basis for running the AutoML job, of which relevant metrics and parameters are automatically recorded and tracked by Mlflow.

3.4 LLM-Assisted AutoML

LLMs assist the user at multiple locations in the *SEDAR* framework. A new user interface based on *LangChain* has been implemented to enable communication between the user and LLM providers, allowing for the dynamic integration of additional information from the system's metadata. The current implementation offers a simple chat functionality from the frontend. The interface can be used with various LLM providers.

JupyterAI is an extension for *Jupyter* that enables the use of LLMs within a separate chat window and the cells of a notebook. It distinguishes between language and embedding models. The embedding model is used to create a compressed representation of the contents of a particular directory, and the language model is used to ask specific questions about the file content. Users can either have methods or code from the notebook explained to them, or have code sections and entire notebooks generated. *JupyterAI* can be connected to a range of default models, but also to *Hugging Face* and *GPT4All*, which include a variety of open-source LLMs of varying sizes.

Furthermore, the *CAAFE* library [9] is used to automate feature engineering of classification problems with tabular data using LLMs (see Listing 1.1).

```
# Iteration 1
# Apply Gaussian blur to images
# Usefulness: Gaussian blur can help smooth out noise in the images
and preserve important edges, which can benefit segmentation tasks
by improving object boundary detection.

import cv2
import numpy as np
X_blurred = np.zeros_like(X)
for i in range(len(X)):
    X_blurred[i] = cv2.GaussianBlur(X[i], (5, 5), 0)[:, :, np.newaxis]
X = X_blurred
Performance before adding features IOU 0.808.
Performance after adding features IOU 0.810.
The code was executed and changes to 'X' were kept.
```

Listing 1.1. Output of the first iteration of the CAAFE algorithm when executed for the image segmentation task (see Table 2). It applies Gaussian blur and finds improvement with the resulting model.

The library has been extended to image data in classification and segmentation problems. *CAAFE* processes a dataset D in an iterative process. An LLM is dynamically prompted to generate code. By executing this code on D, a new dataset D' is created, which contains the feature changes of this iteration. An identical model is trained for both datasets and problem-dependent metrics are measured. These metrics are communicated to the LLM for the next iteration, which receives D' in the next iteration if there is an improvement.

4 Evaluation

We evaluate the proposed framework against problems from real-world use cases at the Institute for Surface Technology of the Niederrhein University of Applied Sciences (HIT). The HIT researches (organic) coatings and paints which play a vital role in safeguarding surfaces. ML is applied in this context for experiment suggestions and data analysis. All datasets, code, and training details for this study have been exported from *SEDAR* and made available online. For detailed information about dataset sizes and class distributions, we refer to the corresponding publication, as we have tried to reproduce the experimental settings accordingly as much as possible. The experiments are performed in the described *Jupyter* environment equipped with a *Nvidia Quadro RTX8000* graphics card, two *Intel Xeon Gold 6230R* processors, and 754 GB RAM. Training for all models is initially limited to one hour. For the *AutoKeras* library, for which no time limit can be defined, a division into 50 models for 50 epochs is used.

Because working with LLMs can be freely designed by the user, rules to improve reproducibility are defined. LLMs in *GPT4All* are designed for the English language, hence the conversations are conducted in English. Problems are described by the user. It is assumed that the user cannot provide an exact definition of the sequence of experiments due to a lack of experience. Hence, ideally, the user does not use any ML terminology when working with LLMs. For example, it should not be mentioned that the class distribution of the data set is unbalanced. Instead, it should be checked whether the contextual information provided by the *SEDAR* AutoML solution is sufficient for the LLM to act accordingly. Any further steps identified by the LLM through continued querying for a more robust ML pipeline are tracked as further effort. The two open-source LLMs that are available in *JupyterAI* through *GPT4All* *nous-hermes-llama2-13b* (Hermes) and *mistral-7b-openorca* (Orca) are used by default, as two representatives with varying parameter sizes and instruction and chat model respectively. Of these, we include the best-performing model in the evaluation. The experiment begins with a single prompt (Fig. 4) to solve the problem in one shot. If this prompt does not create a valid ML pipeline, e.g., due to errors in the created code, the prompt is modified, for example, to split the problem into several prompts. In this case, it is tracked to what extent additional effort is required from the user and how demanding the changes are. If no approach with open-source LLMs leads to success, *GPT4* is used.

```
%%ai chatgpt Generate a Python code snippet for an Image Segmentation pipeline using the libraries Tensorflow or Pytorch and other
appropriate libraries, optimized for GPU usage. The code should handle a dataframe called gitter_trainv1 with columns 'image' and
'label'. 'image' contains the path to the image file and 'label' contains the path to the mask file. The dataset contains 348 images.
All images are 256px in height, 256px in width, in GRAY mode, and saved as png. Focus on a simple yet effective neural network architecture
suitable for beginners. You may use a pre-trained model and fine-tune it on the dataset. Include necessary imports, dataset loading and
preprocessing steps, model architecture, and training loop. The code should be ready to run in a Python interpreter without modifications.
datagen = ImageDataGenerator(rescale=1./255)
image_generator = datagen.flow_from_dataframe(dataframe=gitter_trainv1, x_col='image', y_col='label',
                                              target_size=(256, 256), color_mode='grayscale',
                                              class_mode='input', batch_size=32)
base_model = MobileNetV2(input_shape=(256, 256, 1), include_top=False)

x = base_model.output
x = Conv2D(128, (3, 3), activation='relu', padding='same')(x)
x = UpSampling2D((2, 2))(x)
x = Conv2D(64, (3, 3), activation='relu', padding='same')(x)
x = UpSampling2D((2, 2))(x)
x = Conv2D(1, (3, 3), activation='sigmoid', padding='same')(x)

model = Model(inputs=base_model.input, outputs=x)
model.compile(optimizer=Adam(), loss='binary_crossentropy')
model.fit(image_generator, epochs=10)
```

Fig. 4. Example prompt using *Jupyter Magic* with *GPT*4's response to generate the necessary code for solving the image segmentation task. Shortened for brevity.

4.1 Datasets

The first dataset [11] contains images of four different types of surface defects. The images were collected via Laser-scanning microscopy to obtain height profiles on the surfaces and produce an artificial RGB image by a threefold repetition of the height data. This data is evaluated for **object detection**. While the authors dealt with Few-Shot Object Detection [4], these solutions are not included in the AutoML libraries, and except for *GPT*4, none of the LLMs had information about such models. Hence, we compare to the given *RetinaNet* baseline.

For the same dataset [11] of surface defects optical images are available for **image classification**. Here we limit the number of available training samples per class to simulate a Few-Shot setting.

The next dataset [25] consists of images of scratch tests on surfaces to be evaluated for **image segmentation**. Zhang et al. identify the area where the so-called cross-cut test has removed the surface material. A binary mask with the marked regions is available for each image. They use several image augmentations to increase the amount of training data and make the model more robust. For this dataset, only the original images are available, without augmentations. However, feature engineering using *CAAFE* is performed.

The last dataset comes from [12], where methods for analyzing the resistance of coatings and paints are discussed again. The breakthrough point of a needle on a coating is identified by using sensor data about the movement and the force exerted by the needle on the surface. Using this data, the authors conduct **tabular regression** to find the position of the penetration through the surface.

4.2 Criteria

A series of evaluation criteria are defined to conduct a structured evaluation of how effective the AutoML- & LLM-based approaches are in solving various ML tasks within the SEDAR system. If, for example, the effort is evaluated, more effort corresponds to a less favorable value, which is why an ascending color coding is selected (■■■■■). If the completeness of a solution is being evaluated, the process is reversed; higher values are better (■■■■■).

The performance evaluates the predictive capabilities of the models created by the AutoML solutions. Various metrics are considered, e.g. *Mean Average Precision* (MAP) for object detection, and *Intersection over Union* (IuO) for image segmentation.

Where possible, the values are determined using the built-in functions of the AutoML library. If the appropriate metric is not implemented, this is highlighted negatively and a corresponding implementation is imported.

For efficiency, the time and resources required to experiment cover the process from data selection to the finished model. The time calculation in turn includes a) the preparations for the experiment (t_{prepare}), such as the dialogs, b) the time spent in the source code (t_{code}), for example by adapting function parameters, and c) the computing time required (t_{compute}). The complexity of the configuration of an experiment or the effort required for any changes to the code is measured by the following color-coded scale as *work*: (1) immediate execution, (2) simple, (3) medium, (4) extensive configuration and (5) individualized configuration with external resources.

The amount of prior knowledge and programming skills required for the configuration and changes to the code for a functioning pipeline is rated.

Finally, the initial model created by the AutoML process is assessed. A good starting situation either directly provides a versatile selection of models and parameters that the user can continue to use, or, especially when working with LLMs, provides sufficient hints to improve the designed pipeline.

4.3 Results and Discussion

The results of the experiments (Tables 1, 2, 3 and 4) show that the *AutoML-Wrapper* can considerably simplify creating ML pipelines. The user only has to make minimal entries in the frontend and can still create powerful models, which is particularly advantageous for users without much experience. The *AutoML-Wrapper* can autonomously train and evaluate models in a short time if the appropriate settings are made. The performances in terms of evaluation metrics are competitive compared to the given baselines for both the *AutoMLWrapper* and the LLMs. Especially for the classification of paint defects, the training times were remarkably short, which made it possible to quickly create a selection of prototypes. Although the expected user-friendliness is a strength, it also results in reduced flexibility. The user has less control over specific model parameters and the optimization of these. In addition, the performance of the AutoML-Wrapper is directly dependent on the capabilities of the underlying libraries. In

Table 1. Overview of the experiments related to object detection. Only *AutoGluon* can perform object detection. The open-source models were unable to deal with this task, hence *GPT4* had to be involved. Distribution of train/test sets: 140/36

	$\text{Wrapper}_{\text{gluon}}$	LLM_{gpt4}	[11]
MAP	0.46	0.4	0.458
t_{prepare}(min.)	2	1	
t_{code}(min.)	< 1	10	
t_{compute}(min.)	60	12	
work	■□□□□	■■□□□	
Prior knowledge	■□□□□	■■□□□	
Programming expertise	■□□□□	■■□□□	
Initial model	■■■■□	■■■■■	

Table 2. Overview of the experiments in the segmentation data set. Only *AutoGluon* can perform image segmentation. Distribution of train/val/test sets: 174/35/8

	$\text{Wrapper}_{\text{gluon}}$	LLM_{Orca}	[25]
IoU	0.82	0.91	0.96
t_{prepare}(min.)	3	1	
t_{code}(min.)	3	15	
t_{compute}(min.)	9	20	
work	■■□□□	■■■■□	
Prior knowledge	■■□□□	■■□□□	
Programming expertise	■□□□□	■■□□□	
Initial model	■■■□□	■■■□□	

Table 3. Overview of the image classification experiments with limited training data. *AutoGluon*'s few-shot setting leads to the best results. Distribution of train/test sets: first row 20/79 and second row 80/79

	$\text{Wrapper}_{\text{gluon}}$	$\text{Wrapper}_{\text{gluon}}^{\text{few-shot}}$	$\text{Wrapper}_{\text{keras}}$	$\text{LLM}_{\text{Hermes}}$
F1	0.62	0.87	0.6	0.5
F1	0.85	0.99	0.6	0.55
t_{prepare}(min.)	2	2	2	2
t_{code}(min.)	< 1	<1	15	10
t_{compute}(min.)	< 1	< 1	13	17
work	■□□□□	■□□□□	■■□□□	■■■■□
Prior knowledge	■□□□□	■□□□□	■■□□□	■■■■□
Programming expertise	■□□□□	■□□□□	■■□□□	■■□□□
Initial model	■■■■□	■□□□□	■■■□□	■■■■□

Table 4. Overview of the regression experiments on tabular data. Distribution of train/val/test sets: 204/10/41

	$\text{Wrapper}_{\text{gluon}}$	$\text{Wrapper}_{\text{keras}}$	$\text{Wrapper}_{\text{sklearn}}$	LLM_{Orca}	[12]
Mean-Absolute-Error	8.86	64.68	7.38	53.39	4.58
t_{prepare}(min.)	2	2	2	2	
t_{code}(min.)	< 1	<1	<1	5	
t_{compute}(min.)	< 1	7	60	18	
$work_{\text{code}}$	▣□□□□	▣□□□□	▣□□□□	▣▣□□□	
Prior knowledge	▣□□□□	▣□□□□	▣□□□□	▣□□□□	
Programming expertise	▣□□□□	▣□□□□	▣□□□□	▣□□□□	
Initial model	■■■■□	■■■□□	■■■■■	■■■■□	

summary, the module offers an efficient and user-friendly way to quickly perform experiments that also result in powerful models.

Applying LLMs for code generation and process understanding in ML represents an innovative and promising approach. The open-source models offer the advantages of free availability, easy access to experimentation without direct costs, and special hardware. However, the open-source models' generated answers were often incomplete or lacked precision for a direct application. Many times the code had errors and was not executable. Especially for more complex queries for model creation and optimization, the answers were cut off prematurely or were so erroneous that considerable effort was required from the user. The results indicated that the models have difficulty understanding and adapting to specific requirements and contexts. This led to generic or inappropriate code, especially for specialized tasks such as image segmentation. In contrast, *GPT4* provided a significantly improved experience, generating more precise and relevant code that was better adapted to the problem. The answers were almost complete and pointed to additional aspects that could be used directly as an initial model for further questions to improve the model. For example, the open-source models were unable to utilize the COCO format of the object detection dataset, which was no problem for *GPT4*.

5 Conclusion

The paper discussed the application of AutoML methods and LLMs to enhance the ML capabilities of a data lake. We have presented the Python module *AutoMLWrapper* which integrates various AutoML libraries and deployed it to our data lake system *SEDAR*. In addition, we integrated an LLM interface for code generation to have comprehensive support for data science tasks within the data lake. A thorough evaluation of usability and performance compared both methods with baselines from four real-world use cases, highlighting that both extensions can improve the accessibility and efficiency of data analysis. We have shown a fair comparison of 'classical' AutoML tools and LLMs, although more detailed experiments with more use cases are necessary to derive more general rules for the quality of AutoML and LLMs. The limitations in terms of flexibility and control of the *AutoMLWrapper*, as well as the dependency on the capabilities of the underlying libraries, are important aspects to be considered. The choice between open-source and proprietary models depends on the use case. For projects with a limited budget, feasibility studies, or exploratory purposes open-source models are suitable. Proprietary models offer the necessary performance for more demanding tasks. We are confident that our results can serve as an entry point for future research on automating data science tasks with AutoML tools or LLMs.

Acknowledgements. This work has been sponsored by the German Federal Ministry of Education and Research in the funding program "Forschung an Fachhochschulen", project i^2DACH (grant no. 13FH557KX0).

References

1. Alla, S., Adari, S.K.: What Is MLOps? Apress, Berkeley, CA (2021)
2. Chai, C., Wang, J., Luo, Y., Niu, Z., Li, G.: Data management for machine learning: a survey. IEEE TKDE **35**(5), 4646–4667 (2023)
3. Chen, A., Dohan, D.M., So, D.R.: Evoprompting: Language models for code-level neural architecture search (2023)
4. Chen, T., et al.: Dual-awareness attention for few-shot object detection. CoRR **abs/2102.12152** (2021)
5. Diamantini, C., et al.: A knowledge-based approach to support analytic query answering in semantic data lakes. In: Chiusano, S., Cerquitelli, T., Wrembel, R. (eds.) ADBIS. LNCS, vol. 13389, pp. 179–192. Springer, Cham (2022). https://doi.org/10.1007/978-3-031-15740-0_14
6. Hai, R., et al.: Amalur: data integration meets machine learning. In: 39th IEEE ICDE 2023, Anaheim, CA, USA, April 3-7, 2023, pp. 3729–3739. IEEE (2023)
7. Hai, R., Koutras, C., Quix, C., Jarke, M.: Data lakes: a survey of functions and systems. IEEE TKDE **35**(12), 12571–12590 (2023)
8. Hassan, M.M., et. al.: ChatGPT as your personal data scientist (2023)
9. Hollmann, N., et al.: Large language models for automated data science: Introducing caafe for context-aware automated feature engineering. In: Advances in Neural Information Processing Systems, vol. 36 (2024)
10. Hoseini, S., Ali, A., Shaker, H., Quix, C.: SEDAR: a semantic data reservoir for heterogeneous datasets. In: CIKM 2023, ACM (2023)
11. Hoseini, S., et al.: Automated defect detection for coatings via height profiles obtained by laser-scanning microscopy. MLWA **10**, 100413 (2022)
12. Hoseini, S., et. al.: Coatings intelligence: data-driven automation for chemistry 4.0. In: 2024 IEEE 7th (ICPS), pp. 1–8 (2024)
13. Huang, Q., Vora, J., Liang, P., Leskovec, J.: Benchmarking large language models as AI research agents (2023)
14. Jain, A., et al.: Overview and importance of data quality for machine learning tasks. In: Proceedings of ACM SIGKDD, pp. 3561–3562 (2020)
15. Karmaker, S.K., et al.: Automl to date and beyond: challenges and opportunities. ACM Comput. Surv. **54**(8) (2021)

16. Khan, Y., Zimmermann, A., Jha, A., Gadepally, V., d'Aquin, M., Sahay, R.: One size does not fit all: querying web polystores. IEEE Access **7**, 9598–9617 (2019)
17. Liu, H., Ning, R., Teng, Z., Liu, J., Zhou, Q., Zhang, Y.: Evaluating the logical reasoning ability of ChatGPT and gpt-4. arXiv preprint arXiv:2304.03439 (2023)
18. Liu, S., Gao, C., Li, Y.: Large language model agent for hyper-parameter optimization. arXiv preprint arXiv:2402.01881 (2024)
19. Schelter, S., et. al.: Unit testing data with deequ. In: SIGMOD. ACM (2019)
20. Schlegel, M., Sattler, K.: Extracting provenance of machine learning experiment pipeline artifacts. In: Abelló, A., Vassiliadis, P., Romero, O., Wrembel, R. (eds.) ADBIS. LNCS, vol. 13985, pp. 238–251. Springer, Cham (2023). https://doi.org/10.1007/978-3-031-42914-9_17
21. Tornede, A., et al.: Automl in the age of large language models: Current challenges, future opportunities and risks. arXiv preprint arXiv:2306.08107 (2023)
22. Wei, L., He, Z., Zhao, H., Yao, Q.: Unleashing the power of graph learning through llm-based autonomous agents (2023)
23. Yang, Z., , et. al.: Autommlab: automatically generating deployable models from language instructions for computer vision tasks (2024)
24. Zaharia, M., et al.: Accelerating the machine learning lifecycle with MLflow. IEEE Data Eng. Bull. **41**(4), 39–45 (2018)
25. Zhang, G., et al.: Deep learning-based automated characterization of crosscut tests for coatings via image segmentation. JCTR **19**, 671–683 (2021)
26. Zhang, L., et al.: Mlcopilot: unleashing the power of large language models in solving machine learning tasks. arXiv preprint arXiv:2304.14979 (2023)
27. Zhang, S., Gong, C., Wu, L., Liu, X., Zhou, M.: AutoML-GPT: automatic machine learning with GPT. arXiv preprint arXiv:2305.02499 (2023)
28. Zhao, Y., et al.: Analysis-oriented metadata for data lakes. In: Proceedings of the 25th IDEAS, pp. 194–203. ACM (2021)

Machine Learning

BBQ-Tree – A Decision Tree with Boolean and Quantum Logic Decisions

Alexander Stahl(✉) and Ingo Schmitt

Brandenburg University of Technology, Cottbus, Germany
{stahlale,schmitt}@b-tu.de

Abstract. This study proposes the BBQ-Tree, a new logic-based classifier that combines the two concepts of classical Decision Trees and Quantum-Logic Decision Trees into a generalized model. It thus creates a method that has the power to solve classification problems that incorporate both curved and linear decision boundaries, with a particular focus on interpretability. In addition to the model itself, ways for its efficient training are discussed. Our experimental evaluation demonstrates that our approach is able to produce models that remain compact and provide good insights over trends in data while maintaining an accuracy not worse than Decision Trees alone.

Keywords: quantum logic · classifier · interpretable AI · decision tree

1 Introduction

Classification stands as an important pillar of data mining, whose utility extends across an array of tasks that necessitate the assignment of labels or the organization of data into distinct groups. In this study, we propose a new classification method with improved interpretability, called a BBQ Tree (BBQT). It combines advantages of a classical Decision Tree (DT) with the Quantum Logic Decision Tree (QLDT) from [17] and [18]. In a QLDT, the evaluation of a quantum logic expression on an object o yields a value in $[0, 1]$. For a BBQT we use a final threshold decision on a Commuting Quantum Query Language (CQQL) evaluation function over continuous attribute values based on a mixture of Q-, B- and B½-split nodes. As a result, better interpretability is achieved through Q-splits providing trend information, whereas the accuracy can be improved over QLDTs.

1.1 Classifiers and Limitations

The classification problem is concerned with assigning class labels of predefined classes C to objects $o = (x_1, x_2, \ldots, x_n)$ based on their features x. The goal is to predict the classification result $K(o) \in C$, which is an object's predicted class membership. To this end, it is necessary to use labeled data to learn a mapping

J. Tekli et al. (Eds.): ADBIS 2024, LNCS 14918, pp. 201–214, 2024.
https://doi.org/10.1007/978-3-031-70626-4_14

K that approximates the mostly hidden object-class relationship $C(o)$ as closely as possible. Here, without loss of generality, the focus will lie on binary classifiers, therefore only two possible class labels 0 and 1 shall be considered. As a measure of classifier performance, accuracy is a widely recognized and accepted metric. Given the significance of classification, there exist many methods to produce K that use sets of given labeled training data TR as input. Examples are logistic regression, support vector machines, artificial neural networks, Bayes classifiers, k-nearest neighbor and DTs [4]. Many of these methods are able to achieve high accuracy values in many scenarios. However, accuracy alone is not the only measure to evaluate the usefulness of a classification method. It can also be significantly important for human beings to understand and interpret the decisions that a given classifier makes [3,15]. Examples for scenarios or environments in which the interpretability of the classification decision can be crucial are medical diagnoses, jurisdiction and law enforcement. Explainability, and by extension interpretability (a stricter requirement), fosters trust in the user. Moreover, some legal regulations are even mandating its inclusion [3].

The DT [1,15] is a particularly popular method with good interpretability. DTs are tree structures that split data in each node according to well-chosen split attributes and thresholds. In a DT, every leaf represents a class decision. Thus, the classification of an object o is performed by traversing the path from the root, navigating through the relevant inner nodes, until arriving at a leaf node, which ultimately represents the classification result. All of the paths to the leaves of a class effectively form a Boolean expression in disjunctive normal form. Here, several child paths are combined as disjunctions and the nodes of a path are combined as conjunctions. Therefore DTs are regarded as logic-based classifiers. We assume that such classifiers offer particularly good human interpretability.

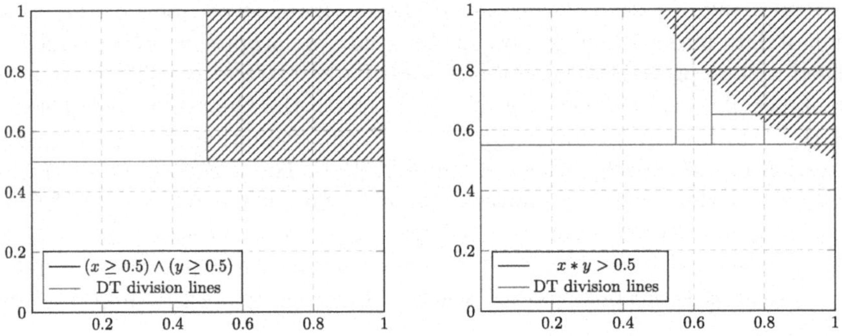

(a) $(x > 0.5) \wedge (y > 0.5)$ (b) $x * y > 0.5$

Fig. 1. Class decision lines of possible DTs for given class shapes

While DTs offer such advantages, they also have some shortcomings. First of all, in a high-dimensional setting with a sufficiently complex DT, the importance of each dimension and the applied local threshold values may appear unintuitive and arbitrary to an observer. Furthermore, the shape of the possible class decision lines may be problematic. Every inner node in a DT, also called an attribute split, represents an axis-parallel cut through the feature space. Thus, classes whose shape in the feature space is irregular (in the sense of non-axis-parallel decision lines) can cause the DT to become deteriorated. Such class decision lines can only be approximated by a DT through a relatively high number of attribute splits. An example for this behavior can be observed in Fig. 1, which depicts a pair of two-dimensional feature spaces, target class shapes and the division lines of a fitting DT. Here the shaded area represents the target class, whereas the red lines visualize the division lines of the DT. The scenario in Fig. 1a shows a case where a DT performs very well, as the shape of the class aligns very well with the axis-parallel class decision lines. In Fig. 1b, however, this is not the case. Here the class shape is defined by the condition $x * y > 0.5$ with $x, y \in [0, 1]$, which manifests in the 2D case as a curve. The axis-parallel class decision lines here can only approximate this shape, which is why a relatively high number of splits per dimension is needed in comparison to the left side. For a DT this means that it will grow relatively deep, which is not desirable for good interpretability. Furthermore, deep tree nodes represent decisions on very fine granularity, which increases the risk of overfitting. There are solutions against overfitting such as pruning and the random-forest approach, which will not be discussed in this study. A QLDT may be able to deal with such class shapes better, using non-parallel class decision lines. This means, however, that a choice between the axis-parallel decision lines of a DT or the non-parallel decision lines of a QLDT has to be made. However, the combination of the two concepts of classical DT and QLDT offers a more general solution. Thus, the goal of the present study is to create a method that has the power to solve classification problems of either nature (linear and curved) with a particular focus on interpretability. This is achieved by creating a model that (a) remains compact while maintaining an accuracy not worse than DT alone and (b) provides good insights over trends in data, as QLDT does.

1.2 Related Works

Qualitative Comparative Analysis and fuzzy set QCA [14] are similar as in that they, too, provide an understanding of the relation between input objects and binary decisions. These approaches work by trying to find conditions for binary decisions that are both necessary and sufficient. They are logic-based, which we already established to be an important property for interpretability. They do not, however, constitute actual classification models on their own.

Another related concept is the Generalized Additive Model (GAM), introduced in [5]. It is a type of statistical model that can depict non-linear relationships between input and output variables. This is achieved through a sum of smooth functions, which capture non-linear patterns in the given data. The GAM is usable in applications such as regression and classification problems. The latter can be achieved by modifying the so-called link function and using appropriate distributions so that categorical outcomes are produced. Furthermore, GAM is considered somewhat interpretable, especially in comparison to black-box models such as neural networks [11]. DTs, however, are inherently interpretable and outperform them in this aspect.

Alongside these approaches, there exists not only the aforementioned DT, but also a multitude of variants that incorporate non-axis-parallel class decision lines. Heath et al. introduced the concept of Oblique Decision Trees [6], which employ linear hyperplanes rather than straightforward attribute thresholds to separate classes within the feature space. Due to the considerable learning effort needed to determine the optimal planes, they opted for a randomized methodology in tree construction. While this approach can yield compact tree structures, the hyperplanes used as splits may lack intuitiveness for human interpretation, and they cannot accommodate curvature.

Instead of assessing just a single feature in a split node, Brodley and Utgoff's Multivariate Decision Tree [2] integrates multiple features, akin to the oblique tree, thus circumventing the constraint of axis-parallel splits. The paper demonstrates that multivariate decision trees exhibit fewer errors on test datasets compared to conventional decision trees. However, the increased complexity of the splits leads to similar challenges in interpretability as observed in oblique decision trees. Moreover, the heightened complexity in learning impacts processes such as feature selection and pruning.

The concept of Non-Linear Decision Tree (NDT) adopts such an approach, but shifts the challenge of crafting more appropriate class decision boundaries away from the classification method itself. Rather than using linear decision lines on the originally given features, it transforms the latter into a new feature space. Using linear decision lines in this new feature space is then equivalent to curved decision lines in the original feature space. The original article by Ittner and Schlosser presents a variety of possible transformations [7]. Yet, it can be argued that such transformations actually decrease the interpretability of the DT, as they are not the product of any logic on the original features. To the human observer the transformations may thus appear arbitrary - their semantic role in the classification can hardly be understood.

Finally, Fuzzy Logic (FL) might possibly yield concepts to provide interpretable classification. However, common FL-derived DT concepts fail to fulfill important logic laws, such as excluded middle or idempotence [8,9,12]. This non-compliance could be seen as yet another barrier to true interpretability. Conversely, CQQL and therefore QLDT do obey the rules of the Boolean algebra.

2 Quantum Logic Decision Tree and Comparison to Classical Decision Tree

In order to extend QLDT, some knowledge about this model is required first. In this section we aim to establish the crucial concepts and motivate all further steps. QLDT was introduced by Schmitt in [17]. A QLDT is a variation of a DT that performs binary classification, using quantum-inspired logic. Notably, this concept applies no techniques of quantum computing. Instead, quantum-inspired logic is used as a means to map logic CQQL [16] expressions to arithmetic expressions. CQQL [16] includes the definition of conditions on attributes of an object. Its syntax is analogous to proposition logic. Unlike fuzzy logic, the evaluation of CQQL conditions obeys the rules of Boolean algebra. All gradual truth values in CQQL fall within the interval $[0, 1] \subset \mathbb{R}$ with 1 being equivalent to `true` and 0 equivalent to `false`. As the input, we assume values of n standardized attributes of an object o. Let then $a^o \in [0, 1]$ be the value of attribute a of object o, where 1 expresses the highest value and 0 expresses the lowest value. It is assumed that all attributes on which conditions can be formulated are probabilistically independent. Note that this does not mean that the method can exclusively operate on independent features. Instead, the dependence among features may manifest itself within the resulting tree structure, should it exist. The evaluation of a CQQL expression will then also return a value out of $[0, 1]$. Let, for example, a, a_1, a_2 be several attributes and $[\cdot]^o \in [0, 1]$ denote the evaluation of a CQQL condition on an object o. The following example mappings (which do not constitute general rules!) of expressions over the attributes a, a_1, a_2 demonstrate the concept of CQQL mapping. For further details please refer to [16].

- $[a]^o = a^o$ (attribute value)
- $[\neg a]^o = 1 - a^o$ (negation)
- $[a_1 \wedge a_2]^o = a_1^o * a_2^o$ (conjunction)
- $[a_1 \vee a_2]^o = a_1^o + a_2^o - a_1^o * a_2^o$ (disjunction)

Before an arbitrary expression is evaluated, it needs to be transformed into disjunctive normal form, which can be achieved with standard Boolean algebra rules. For the purpose of classification, a suitable classifier K has to be derived from training data TR. Here, K will consist of a CQQL expression e as well as a well-chosen output threshold $\tau \in [0, 1]$. For classification, the numeric result of the evaluation of e is calculated and the threshold τ applied:

$$K_e^\tau(o) = th_\tau([e]^o) \text{ with } th_\tau(x) = \begin{cases} 1 & \text{if } x > \tau \\ 0 & \text{otherwise.} \end{cases} \tag{1}$$

As it was presented in [17], the best expression $e = \bigvee_{i \in I} m_i$ (disjunction over minterms) is found by determining for every possible complete minterm m_i, whether it should be active $i \in I$ or inactive $i \notin I$ for a high classification accuracy value. For this purpose, we use the following rule:

$$m_i \text{ is } \begin{cases} \texttt{active} & \text{if } \rho_i > \rho_\theta \\ \texttt{inactive} & \textit{otherwise} \end{cases} \text{ with } \rho_i = \frac{O_i}{(O_i + Z_i)} \tag{2}$$

where $\rho_\theta \in [0, 1]$ is a well-chosen activation threshold for single minterms, which is to be regarded as a hyperparameter of the QLDT. Furthermore, O_i and Z_i are defined as $O_i = \frac{1}{n_O} * \sum_{(o,1) \in TR} \prod_{j=1}^{n} [c_{ij}]^o$ and $Z_i = \frac{1}{n_Z} * \sum_{(o,0) \in TR} \prod_{j=1}^{n} [c_{ij}]^o$ with

$n_O = |\{(o, 1) \in TR\}|$ and $n_Z = |\{(o, 0) \in TR\}|$ being the number of 1- and 0-objects in *TR* respectively. Here the term c_{ij} refers to the j-th attribute condition of minterm i (non-negated $[c_{ij}]^o := [a_j]^o = a_j^o$ or negated $[c_{ij}]^o := [\neg a_j]^o = 1 - a_j^o$). The formula $\prod_{j=1}^{n}[c_{ij}]^o$ is the conjunctive evaluation (i.e. the product) of a complete minterm m_i for an object o. Thus, for a given minterm m_i and *TR*, the decision is made, whether an average 1-object minterm score is higher than the average 0-object minterm score. A complete minterm corresponds to a corner of a hypercube $[0, 1]^n$ and the index set I of e corresponds to a set of active corners expressing the trend, that there reside more 1-objects than 0-objects. Having found a suitable expression e from *TR*, the approach presented in [17] can then convert e into a quantum logic decision tree $qldt(e)$. The tree-based representation serves to enhance interpretability. After the tree is constructed, its arithmetic evaluation for a given object o is performed by summing up over the evaluations of all paths to its 1-leaves L_c. Here, different paths correspond to exclusive disjunctions. The conditions of one path, however, correspond to conjunctions, leading to:

$$[e]^o = [qldt(e)]^o = \sum_{i \in L_c} \prod_{j \in path_i} [c_{ij}]^o. \tag{3}$$

For the determination of a suitable threshold τ, let $\min_O = \min_{(o,1) \in TR}([e]^o)$ be the smallest evaluation result of the 1-objects and $\max_Z = \max_{(o,0) \in TR}([e]^o)$ the highest result of the 0-objects from *TR*. In case of $\max_Z < \min_O$, 1-objects and 0-objects are well separated and we set τ to $(\max_Z + \min_O)/2$. Otherwise, we choose an appropriate value τ from the interval $[\min_O, \max_Z]$. In order to find such a threshold which maximizes accuracy we also utilize the training objects from *TR*:

$$\tau = \underset{\tau_o \in [\min_O, \max_Z]}{\arg\max} \left(accuracy\,(e, \tau_o, TR)\right) \tag{4}$$

Here it might be helpful to compare QLDTs with DTs to gain a full understanding of the concept. In this scope, we consider only DTs that work with attribute thresholds. In a classical DT, the evaluation of a path $path_i$ to a leaf i corresponds to a Boolean conjunction $\bigwedge_{j \in path_i}[c_{ij}]^o$ ($[c_{ij}]^o \in \{0, 1\}$) and can be arithmetically expressed by $\prod_{j \in path_i}[c_{ij}]^o \in \{0, 1\}$, whereas the QLDT evaluation is given analogously by $\prod_{j \in path_i}[c_{ij}]^o \in [0, 1]$. Then, the evaluation of an object o in a classical DT is given by $\prod_{j \in path_i}[c_{ij}]^o = 1$ for exactly one leaf i, whereas for a QLDT, more than one leaf evaluation can be non-zero. Therefore, the evaluation is performed by summing up all 1-leaf evaluations. This also provides a generalization of the classical DT, since all paths to leaves that do not correspond to the object's class are zero. Finally, in a classical DT, a path to a leaf can contain several conditions on the same attribute, whereas in a QLDT every attribute occurs at most once per path. An example QLDT with its equivalent CQQL expression can be found in Fig. 2, which also provides an optimal solution for the classification problem in Fig. 1b (given $\tau = 0.5$). Note that the notation chosen in this study depicts the non-negated path to a child node as a solid line and the negated path as a dashed line. Class labels of the value 0 may be omitted.

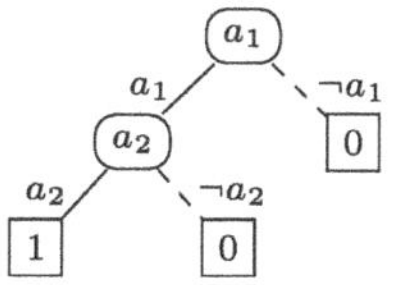

$$e = a_1 \wedge a_2$$

$$[e]^o = [qldt(e)]^o = a_1^o * a_2^o$$

Fig. 2. An exemplary QLDT, representing $a_1 * a_2 > \tau$

The global search for a suitable QLDT is computationally expensive. Thus, Schmitt developed the greedy method QLDT+ to create a QLDT $qldt^e$ in a stepwise manner from the root to the leaves analogously to classical DT construction, that is, not every of the 2^n minterms needs to be considered in advance as it is the case for applying $qldt(e)$. For the construction of a QLDT, a node decision is made recursively until a leaf is reached in all subtrees. For details on the algorithm, please refer to [18].

3 BQ-Tree: B½-split and Q-splits

In this section, we integrate the formerly separate concepts of DT and QLDT. From now on, we will refer to QLDT attribute conditions as Q-splits, with "Q" referring to their "quantum" evaluation. In order to make DTs compatible to QLDT, we are forced to apply the following restrictions.

As mentioned before, we will only consider binary DT classifiers. Moreover, all n object attribute values are continuous and normalized: $o \in [0,1]^n$. Instead of considering arbitrary thresholds for the DT splits, we define a fixed threshold at ½. A classical DT split with such a threshold at ½ will be called B½-split, where "B" refers to "Boolean". Thus, we can define a combination of QLDT and DT, called a BQ Tree (BQT). Under the defined restrictions, its evaluation is performed analogous to Eq. (3) and $[c_{ij}]$ as follows: Let a be the corresponding attribute of c_{ij} and $a^o \in [0,1]$ the attribute value of object o. Then, the evaluation $[c_{ij}]^o$ can be seen in Eq. (5) below, where $th = 0.5$ holds for B½-splits.

$$[c_{ij}]^o = \begin{cases} a^o \text{ on non-negated Q-split:} [c_{ij}]^o \hat{=} [a]^o \\ 1 - a^o \text{ on negated Q-split:} [c_{ij}]^o \hat{=} [\neg a]^o \\ 1 \text{ on non-negated B-split:} [c_{ij}]^o \hat{=} [a > th]^o \text{ and } a^o > th \\ 0 \text{ on non-negated B-split:} [c_{ij}]^o \hat{=} [a > th]^o \text{ and } a^o \leq th \\ 0 \text{ on negated B-split:} [c_{ij}]^o \hat{=} [\neg(a > th)]^o \text{ and } a^o > th \\ 1 \text{ on negated B-split:} [c_{ij}]^o \hat{=} [\neg(a > th)]^o \text{ and } a^o \leq th \end{cases} \quad (5)$$

For enhanced insight, in Fig. 3 we include a visual representation of the effects of Q-splits and B½-splits. Figure 3a shows a visualization of a decision scenario for a given BQ-node at a path i, given TR and with a new attribute a to be checked. The figure depicts training objects as blue or red dots with attached lines, representing 0-objects and 1-objects, respectively. The x-axis refers to the attribute value $a^o \in [0,1]$, while the y-axis refers to the score of the path $[path_i]^o$ from the root to the node i. The red dashed line stands for the average path

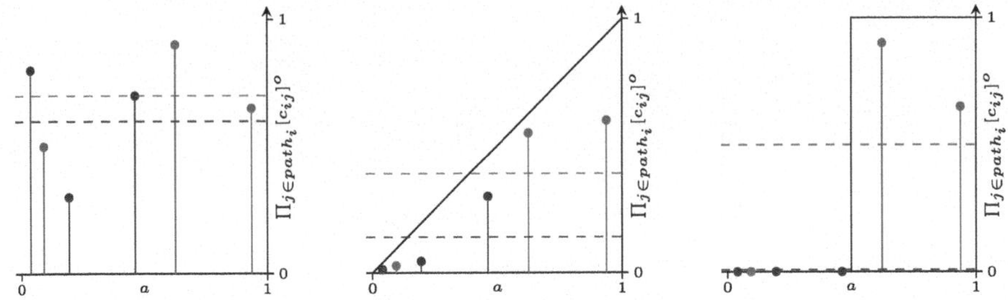

(a) Values of a ($O = 0.68, Z = 0.6, \rho = 0.53$)

(b) Q-split ($O = 0.39, Z = 0.14, \rho = 0.73$)

(c) B1/2-split ($O = 0.52, Z = 0, \rho = 1$)

Fig. 3. Effects of BQ-splits at path i on values of attribute a

score O of all 1-objects, whereas the blue one stands for the average path score Z of all 0-objects. If we were to perform a leaf decision for this scenario with a given $\rho = 0.53$, we would conclude that this depicted node is very far away from being a leaf, due to the red and blue lines being too close to one another. The optimal split with regard to Fig. 3b would be the one that separates the red and blue dashed lines as much as possible. This corresponds to our split criteria, where we choose the split that maximizes *diff*.

Figure 3b depicts the same kind of visualization for the application of a Q-split on the attribute a. Here the effect of the split on O, Z and ρ is shown, which corresponds to a multiplication of the path score with a^o. This can also be interpreted as giving the score a weight with respect to the corresponding attribute value. It is visible that the left split is the optimal one, as it creates a higher separation between the dotted red and blue lines. Finally, Fig. 3c shows the effect of the application of a B1/2-split. It can be seen that the scores are either completely unaffected (i.e. multiplied by 1) or reduced to 0. The black box/line further visualizes this effect by splitting the domain at $a = 0.5$. Once again, the optimal split here would be the one with the highest separation between the average O and Z.

With deepened understanding of the relative effects of Q-splits and the newly introduced B1/2-splits, we are able to construct a BQT using these split types. An example for a simple BQT can be seen in Fig. 4. On the left side of the figure, the tree itself is depicted, whereas on the right side, the evaluation formula and a visualization of the class separation lines for $\tau = 0.5$ are given.

Finally, another alternative of using the concept should be mentioned, aside from the employment of the BQT directly for classifications. Since the BQT works by generating a score that is subjected to an output threshold function, it is also possible to use this score indirectly. A possible application of the score is to use it as an additional feature (attribute) for other classification methods, most notably the DT. In such a scenario, all original available features as well as the score of a BQT are used to train a DT. Section 5 will assess the effectiveness

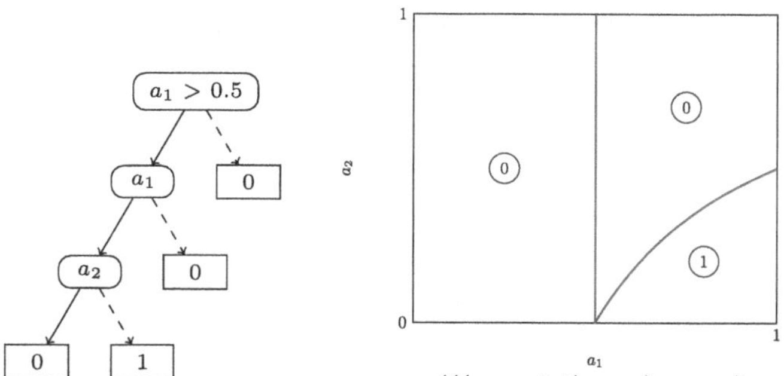

$$e = (((a_1 > 0.5) \wedge a_1) \wedge \neg a_2)$$
$$[e]^o = [a_1^o > 0.5]^o * a_1^o * (1 - a_2^o)$$

Fig. 4. A simple BQT with one $B^{1/2}$-split and two Q-splits and $\tau = 0.5$

of both this approach and direct classification using BQT, including its extension to BBQT.

4 BQT Construction and Transformation Into BBQT

Constructing a BQT from training data can be achieved by an extension of QLDT+ [18], which was mentioned in Sect. 2. As stated there, the two main decisions to make are (a) whether the current node will be a leaf and, if not, (b) the specifics of the split. In the QLDT+ algorithm, a leaf is reached if ρ (see Eq. (2)) is small enough for a 0-leaf or high enough for a 1-leaf or if there no more remaining attributes. Calculating ρ remains possible for both Q- and $B^{1/2}$-splits, which is why this leaf decision method can be adopted for BQT. As for the split attribute decision, the algorithm employs ρ as a difference $diff = |\rho_{\nu a} - \rho_{\nu \overline{a}}|$ for the best purity in node ν. For the choice of a split dimension in BQT, we can adopt this method as well. Additionally, since splits can be of different types, it must also be decided whether a split is a Q- or a $B^{1/2}$-split. To compare each split for a given node, we may also use ρ analogously to the split attribute decision.

To further increase the flexibility of the model, we define a balance factor $bq\_balance \in [0, 1]$ as a hyperparameter. This allows us to control the decision between Q- and $B^{1/2}$-splits by assigning different weights to them. For a $B^{1/2}$-split, the ρ-difference is thus calculated as $bal\_diff = bq\_balance * diff$ and for a Q-split as $bal\_diff = (1 - bq\_balance) * diff$. The split with the highest bal_diff value is then taken as the optimal split.

A remaining open question is, how attribute splits with the best possible local threshold can be achieved, as they are known from DT. To resolve this issue, we define another split type that corresponds to a DT split with a local threshold $th \in [0, 1]$ that is not limited to the value of 0.5. We call this type of split a B-split and a tree that contains all of these splits a BBQT. With the new split incorporated, the evaluation $[c_{ij}]^o$ from Eq. (5) can be used for arbitrary th values.

Experiments with the split attribute decision show that B-splits are not directly compatible with B½-splits and Q-splits. The latter two contain trend information about the boundaries of $[0,1]^n$ and are used for improved interpretability and explainability. On the other hand, the addition of B-splits primarily aims to further enhance the model's classification accuracy. In the learning process, this creates certain challenges. In local split decisions that are used by greedy learning algorithms, B-splits generally hold an advantage with regards to being chosen because they offer an immediate separation effect on their own and are relatively powerful due to their adaptable local threshold. This effect can be observed in the QLDT+ algorithm as well. Q-splits, on the other hand, work better through the combination with other Q-splits of their corresponding path. Due to this incompability, it is not possible to consider B-splits during the initial training of the BBQT. Instead, we separate the training into two phases: In the first phase, a BQT is constructed as before. To avoid overfitting and to achieve simple and interpretable trees, the size of a BQT can also be restricted. After that, it may be refined by the addition of B-splits, following traditional DT split decisions. This second step transforms a BQT into a BBQT.

Note that integrating B-splits into the tree not only enhances its accuracy but also conceptually enables the utilization of categorical attributes. For this case, a B-split would check for equality to a suitable value instead of the threshold comparison. QLDT and BQT were not designed to handle categorical data without the usage of encoding methods that map categories to $[0,1]$. This expansion in functionality broadens the applicability of BBQT compared to those tree types.

In terms of interpretability, we hold that both QLDT and BBQT offer it to a greater extent compared to standard DTs and their variants. While it may not be possible to prove interpretability formally, several points support this assertion. Firstly, the Q-splits in BBQTs encapsulate trend information rather than relying solely on threshold values, as in DTs. While local thresholds in DT splits effectively serve the classification purpose, they lack clarity for human observers regarding the significance of the used values in particular. In contrast, a Q-split simply indicates whether a feature should be considered 'high' or 'low' to yield a certain outcome, which can be deemed more intuitive. Secondly, through the adherence to the CQQL normal form, BBQTs ensure that each path contains at most one Q-split per feature. This constraint prevents the proliferation of complex and lengthy paths within the tree. Furthermore, BBQTs that were generated via QLDT+ additionally are ordered in such a way that more significant splits are located closer to the root. This simplifies the fast interpretation of the tree's logic. Finally, by imposing limitations on tree height and enabling both linear and non-linear class separation, the size of the tree can be minimized compared to traditional DTs, further enhancing interpretability.

5 Experimental Study

To validate the performance of our new approach, we conducted experiments using a publicly available dataset called "Banknote Authentication". The experiments apply our methodology to the dataset and evaluate and compare the resulting models with regards to their accuracy and interpretability. We desire a compact logical formula that provides trend information about in which corners or boundaries of the unit cube the 1-objects mainly reside. In the following, we will give a description of the data used as well as the results gained. For the determination of accuracy values, we performed k-fold cross validation with $k = 10$. Please note that the trees featured in this section are examples from the given configurations that were chosen for illustrative purposes. The used input parameters were chosen empirically for the algorithm, based on observed effectiveness. Our evaluation involves the attained accuracy and tree size for the following options: classical DT, BQT, BBQT and the combination of DT and BQT. DTs were generated with scikit-learn [13].

The used dataset contains measurements that determine whether a banknote is valid (1) or not (0), which, for example, may be the case for a counterfeit. For purposes of validation, this dataset may be accessed online via [10]. In it, there are four continuous attributes, which were extracted from photographs of the banknotes. The feature `entropy` (e) was extracted from the original image, whereas `variance` (v), `kurtosis` (k) and `skewness` (s) were generated from its wavelet transformation. The data is comprised of 1220 tuples. For each experiment, the training data was balanced via random undersampling. The results of the application of various tree classifiers, including BQT and BBQT as well as their respective training configuration, are shown in Table 1. Here, the *level* column refers to the maximum tree height. For BBQT it is given as two values $l_1 + l_2$, which means that first a BQT was trained with at most l_1 levels and then turned into a BBQT by attaching B-splits of at most l_2 additional levels. The tree type "BQT+DT" refers to a DT which uses the score of a BQT as an additional feature. For these cases, l_1/l_2 denote the levels of the DT/BQT respectively.

Table 1. Results for banknote validation dataset

#	Tree Type	Level	ρ_θ	*bq_balance*	Acc.
1	DT	5	-	-	97
2	QLDT	4	0.8	0.1	93
3	QLDT	3	0.6	0.1	92
4	BQT	3	0.6	0.3	89
5	BQT	4	0.5	0.3	90
6	BBQT	3+2	0.5	0.1	97
7	BBQT	3+2	0.5	0.3	97
8	BBQT	2+3	0.5	0.1	96
9	BBQT	5+0	0.5	0.3	96
10	BQT+DT	3/3	0.6	0.3	96
11	BQT+DT	4/3	0.6	0.3	97

A BQT trained on the dataset (configuration #4) can be seen in Fig. 5a and for comparison, a standard DT in Fig. 5b. Evaluating this BQT, two disjunctive boundaries of $[0, 1]^4$ are noticeable:

1. $\overline{\texttt{variance}} \land \texttt{skewness} \land (\texttt{kurtosis < 0.5})$
2. $\overline{\texttt{variance}} \land \overline{\texttt{skewness}}$

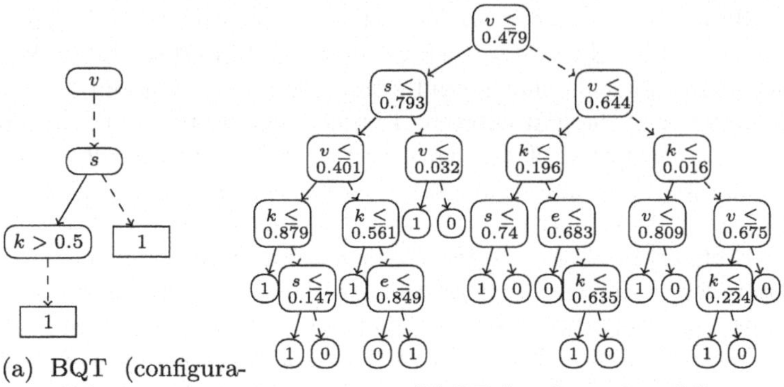

(a) BQT (configuration #4)

(b) DT (configuration #1)

Fig. 5. BQT and DT for banknote classification

Figure 6 shows a BBQT for the banknote classification dataset, that was generated by configuration # 7 in Table 1. In its upper levels, a similar structure to tree #4 (regarding the selection and order of splits) can be observed. Further towards the bottom, B-splits were attached that give this tree an improved accuracy but are less important for the interpretation of results than upper levels. We also tested the training of a DT with a BQT feature. The result is found as #11 in Table 1. This was done by adding a previously trained BQT's score value to the existing training data as an additional feature. In our experiments, the BQT feature was selected as the first split node, meaning it is the best possible feature to base a class decision on.

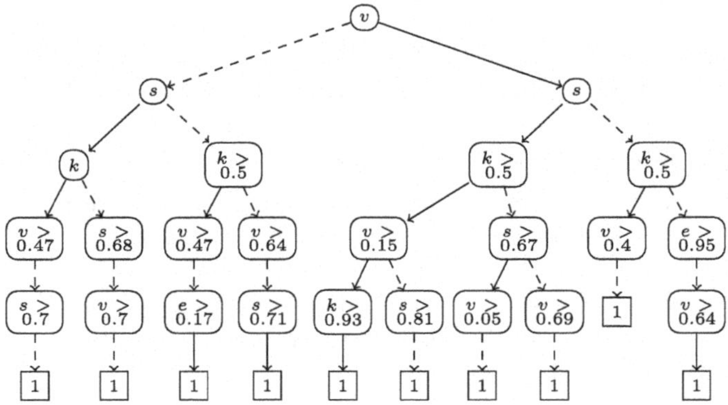

Fig. 6. BBQT for banknote classification with 97% accuracy (configuration #7)

Presented in Table 1 is a summary of the results for the dataset. It was noticeable that the BQT alone showed no improvement of accuracy over the classical DT. Instead, its true advantage appears to lie in its relative compactness (which leads to good interpretability) while retaining a roughly similar accuracy. The BBQT, however, does attain comparable accuracy values. In general, BQT and BBQT appear to be easier to interpret and allow for conclusions about trends in the data. As for the case of BQT+DT, the addition of another feature merely give the DT another option. This effectively means that doing so cannot decrease the accuracy on training data, as the DT may still only choose the original features if they offer higher purity. In other words, adding a BQT feature can either improve or not influence the accuracy of a DT, but should not worsen it. Using BQT in combination with DT also seems to lead to more compact DT. Of course it should be noted that for full interpretability of such a DT, the extra BQT will have to be taken into account. However, we contend that the interpretability of this model surpasses that of a comparable NDT, where any extra feature is only determined by good accuracy. In our combined case, however, the extra feature can be understood completely due to the BQT present.

6 Conclusion

In the present study, we combined the concepts of Quantum Logic Decision Trees (QLDT) and Decision Trees (DT) to create a new classification model called BBQ-Tree (BBQT) and its precursor, the BQ-Tree (BQT). QLDT is an alternative to DT that offers more interpretable classification through the usage of logic and trend evaluation. Through the global evaluation of a QLDT on an object, a score is calculated, on which a well-chosen threshold is applied to perform binary class decisions. For our concept, we introduced another type of split, the B-split and its special case, the B½-split. It is these new split types, in combination to the already known Q-split from QLDT, which form the name for this new tree. Our training algorithm, based on QLDT+, first produces a BQT, i.e. a QLDT with B½-splits. In a subsequent postprocessing step the accuracy can then be increased further by introducing B-splits to the BQT, creating a BBQT. Finally we also explored a variant of a DT that uses a trained BQT score as a feature for its own training.

In order to illustrate the usefulness of our approach, we performed experiments on a test dataset, evaluating the trees' developments of accuracy, compactness and interpretability. Of the introduced concepts, the BQT alone usually offers no accuracy improvement over a DT in the considered scenario, but already yields better interpretability. BBQTs were able to match the accuracy of DT while maintaining the improved interpretability of QLDTs. DTs that were trained with a BQT feature generally performed best (or at least as well as DTs).

Generally, BQTs and BBQTs are able to offer more insights, as they combine the evaluation of trends in data with logic. As for the DT with BQ features, the performance can never be worse for training data, since the DT training algorithm still has access to all original features and may select them if they provide the best results. Thus, the combination can only be either as good as the normal DT or better, and tends to lead to more compact DTs with higher interpretability than similar approaches.

Disclosure of Interests. The authors have no competing interests to declare that are relevant to the content of this article.

References

1. Breiman, L., Friedman, J.H., Olshen, R.A., Stone, C.J.: Classification and Regression Trees. CRC Press, New York (1984). https://doi.org/10.1201/9781315139470
2. Brodley, C.E., Utgoff, P.E.: Multivariate decision trees. Mach. Learn. **19**, 45–77 (1995)
3. Doshi-Velez, F., Kim, B.: Towards a rigorous science of interpretable machine learning (2017). https://doi.org/10.48550/ARXIV.1702.08608
4. Han, J., Kamber, M., Pei, J.: Data Mining: Concepts and Techniques, 3rd edn. The Morgan Kaufmann Series in Data Management Systems, Boston (2012). https://doi.org/10.1016/C2009-0-61819-5
5. Hastie, T., Tibshirani, R.: Generalized additive models. Stat. Sci. **1**(3), 297–310 (1986)
6. Heath, D., Kasif, S., Salzberg, S.: Induction of oblique decision trees. In: IJCAI, vol. 1993, pp. 1002–1007. Citeseer (1993)
7. Ittner, A., Schlosser, M.: Non-linear decision trees - NDT. In: International Conference on Machine Learning (1996)
8. Janikow, C.Z.: Fuzzy decision trees: issues and methods. IEEE Trans. Syst. Man Cybern. Part B Cybern. **28**(1), 1–14 (1998)
9. Jiménez, F., Martínez, C., Marzano, E., Palma, J.T., Sánchez, G., Sciavicco, G.: Multiobjective evolutionary feature selection for fuzzy classification. IEEE Trans. Fuzzy Syst. **27**, 1085–1099 (2019)
10. Lohweg, V.: Banknote authentication. UCI Machine Learning Repository (2013). https://doi.org/10.24432/C55P57
11. Lou, Y., Caruana, R., Gehrke, J.: Intelligible models for classification and regression. In: Proceedings of the 18th ACM SIGKDD International Conference on Knowledge Discovery and Data Mining, KDD 2012, pp. 150–158. Association for Computing Machinery, New York (2012). https://doi.org/10.1145/2339530.2339556
12. Olaru, C., Wehenkel, L.: A complete fuzzy decision tree technique. Fuzzy Sets Syst. **138**(2), 221–254 (2003). https://doi.org/10.1016/S0165-0114(03)00089-7
13. Pedregosa, F., et al.: Scikit-learn: machine learning in Python. J. Mach. Learn. Res. **12**, 2825–2830 (2011)
14. Rihoux, B., Ragin, C.C.: Configurational Comparative Methods: Qualitative Comparative Analysis (QCA) and Related Techniques. Sage Publications, Thousand Oaks (2009). https://doi.org/10.4135/9781452226569
15. Russell, S., Norvig, P.: Artificial Intelligence: A Modern Approach, Global Pearson Education, London (2021)
16. Schmitt, I.: QQL: a DB&IR query language. VLDB J. **17**(1), 39–56 (2008). https://doi.org/10.1007/s00778-007-0070-1
17. Schmitt, I.: QLDT: a decision tree based on quantum logic. In: Chiusano, S., et al. (eds.) New Trends in Database and Information Systems, pp. 299–308. Springer, Cham (2022). https://doi.org/10.1007/978-3-031-15743-1_28
18. Schmitt, I.: QLDT+: efficient construction of a quantum logic decision tree. In: Proceedings of the 8th International Conference on Machine Learning Technologies, ICMLT 2023, Stockholm, Sweden, 10–12 March 2023, pp. 82–88. ACM (2023). https://doi.org/10.1145/3589883.3589895

Large Language Models

Using LLMs for the Extraction and Normalization of Product Attribute Values

Alexander Brinkmann(✉), Nick Baumann, and Christian Bizer

University of Mannheim, Schloss, 68161 Mannheim, Germany
{alexander.brinkmann,christian.bizer}@uni-mannheim.de,
nick.baumann@students.uni-mannheim.de

Abstract. Product offers on e-commerce websites often consist of a product title and a textual product description. In order to enable features such as faceted product search or to generate product comparison tables, it is necessary to extract structured attribute-value pairs from the unstructured product titles and descriptions and to normalize the extracted values to a single, unified scale for each attribute. This paper explores the potential of using large language models (LLMs), such as GPT-3.5 and GPT-4, to extract and normalize attribute values from product titles and descriptions. We experiment with different zero-shot and few-shot prompt templates for instructing LLMs to extract and normalize attribute-value pairs. We introduce the Web Data Commons - Product Attribute Value Extraction (WDC-PAVE) benchmark dataset for our experiments. WDC-PAVE consists of product offers from 59 different websites which provide schema.org annotations. The offers belong to five different product categories, each with a specific set of attributes. The dataset provides manually verified attribute-value pairs in two forms: (i) directly extracted values and (ii) normalized attribute values. The normalization of the attribute values requires systems to perform the following types of operations: name expansion, generalization, unit of measurement conversion, and string wrangling. Our experiments demonstrate that GPT-4 outperforms the PLM-based extraction methods SU-OpenTag, AVEQA, and MAVEQA by 10%, achieving an F1-score of 91%. For the extraction and normalization of product attribute values, GPT-4 achieves a similar performance to the extraction scenario, while being particularly strong at string wrangling and name expansion.

Keywords: Information Extraction · Product Attribute Value Extraction · Value Normalization · Large Language Models

1 Introduction

Product attribute value extraction (PAVE) identifies attribute values in product titles and descriptions. After normalizing the extracted attribute values to attribute-specific scales, they are used for tasks such as faceted product search or

J. Tekli et al. (Eds.): ADBIS 2024, LNCS 14918, pp. 217–230, 2024.
https://doi.org/10.1007/978-3-031-70626-4_15

product comparison. Figure 1 shows an example of a product offer and attribute-value pairs that have been extracted from the product title. For each attribute, the extracted and the normalized value are displayed.

Fig. 1. Product offer with extracted and normalized attribute-value pairs.

Existing methods for PAVE often require large amounts of domain-specific training data to fine-tune pre-trained language models (PLM) that label attribute value sequences [9,24,29] or extract attribute values using question answering [21,26]. The methods focus on identifying the sequences of tokens that form attribute values but do not cover the normalization of the extracted values. Motivated by the success of large language models (LLMs) in related NLP tasks [22] and other information extraction use cases [1,3,12], this paper explores the potential of LLMs for the following PAVE tasks: (i) direct extraction, (ii) extraction with normalization, and (iii) normalization. The objective of the direct extraction task is to extract sequences of tokens that form attribute values from product titles and descriptions [21,23,24,26]. The goal of the extraction with normalization task is to extract and normalize attribute-value pairs in a single step. The normalization task aims at normalizing attribute values that were extracted by a separate, preceding extraction step [8]. Since current benchmarks for extracting product attribute values are designed for measuring extraction quality and do not cover the normalization of attribute values [23,26,28], we introduce a new benchmark dataset, Web Data Commons - Product Attribute Value Extraction (WDC-PAVE). Unlike existing benchmark datasets, which contain only data from a single source [23,26,28], WDC-PAVE consists of 565 product offers originating from 59 different websites that use the schema.org vocabulary. In contrast to related work [23,26], the 4,687 attribute-value pairs in WDC-PAVE have been manually verified and are available in two formats: (i) extracted and (ii) normalized. To normalize the attribute-values, the systems need to perform name expansion, generalization, unit of measurement conversion, and string wrangling. In summary, this paper makes the following contributions:

1. We propose prompt templates for instructing LLMs to extract and normalize attribute-value pairs from product titles and descriptions. The templates cover use cases with and without training data. In contrast to existing work on attribute value normalization [8], the templates exploit the attribute value context for the normalization.

2. We introduce the WDC-PAVE benchmark consisting of 565 heterogeneous product offers and 4,687 manually verified attribute-value pairs. The benchmark supports three tasks: (i) direct extraction, (ii) extraction with normalization, and (iii) normalization. In contrast to existing benchmarks [23,26,28], WDC-PAVE covers value extraction and value normalization.
3. We experimentally compare the extraction performance of GPT-3.5 and GPT-4 to the PLM-based extraction methods SU-OpenTag [23], AVEQA [21], and MAVEQA [26] on WDC-PAVE. GPT-4 achieves the overall best results with an F1-score of 91% and outperforms the PLM baselines by 10%.
4. We experiment with extracting and normalizing attribute values in a single step using GPT-3.5 and GPT-4. Given 10 example attribute values and 5 demonstrations, GPT-4 again reaches and overall F1-score of 91%. The model performs particularly well for string wrangling and name expansion with F1-scores of 95% and 98% respectively.
5. We experiment with normalizing previously extracted attribute values using GPT-3.5 and GPT-4. Given 10 example attribute values and 5 demonstrations, GPT-4 reaches a F1-score of 96%, which is 5% higher than in the extraction and normalization scenario.

The paper is structured as follows: Section 2 introduces the benchmark dataset WDC-PAVE. Section 3 describes the experimental setup. Section 4, Sect. 5, and Sect. 6 discuss the experimental results for the scenarios: (i) direct extraction, (ii) extraction with normalization, and (iii) normalization of product attribute values. Related work in discussed in Sect. 7. The WDC-PAVE benchmark and the code for replicating the experiments are available online[1].

2 The WDC-PAVE Benchmark

This section introduces the WDC-PAVE benchmark. First, we describe the collection of product offers and attribute-value pairs using schema.org[2] annotations and product specification tables within web pages. Second, we present profiling statistics about the WDC-PAVE benchmark. Third, we introduce the normalization operations.

Data Collection. The Web Data Commons (WDC)[3] project extracts structured data from the Common Crawl[4] and provides the extracted data for public download. The WDC Product Data Corpus (WDC LSPM)[5] [13] is one of the extracted datasets. It consists of over 26 million product offers originating from 79 thousand different websites which employ the schema.org vocabulary to annotate structured product data within their HTML pages. The offers are classified into 26 product categories. In addition to schema.org annotations, WDC LSPM

[1] https://github.com/wbsg-uni-mannheim/wdc-pave.
[2] https://schema.org/.
[3] https://webdatacommons.org/.
[4] https://commoncrawl.org/.
[5] https://webdatacommons.org/largescaleproductcorpus/v2/.

extracts attribute-value pairs from specification tables found in the web pages. The attributes in these pairs are product category-specific, such as the number of processor cores of a computer. Category-specific attributes are not part of the schema.org vocabulary and therefore are not explicitly annotated in the web pages. We clean the product offers and attribute-value pairs in the WDC LSPM corpus, omitting those with missing titles, descriptions, or specification tables, and those with descriptions exceeding 1,000 characters. In addition, HTML and language tags are stripped away, and only product offers in English are kept. We select the five categories 'Computers and Accessories', 'Jewelry', 'Grocery and Gourmet Food', 'Office Products', and 'Home and Garden' for WDC-PAVE because they contain a large number of product offers and attribute-value pairs after pre-processing. Subsequently, a random sample of product offers is drawn for each category, with the objective to manually verify their attribute-value pairs. Based on the sampled product offers a fixed set of attributes per category is determined. As the attribute-value pairs in the specification tables are heterogeneously annotated on the different websites, a human annotator is required to verify that each attribute value is a sub-string of the title or the description, and that it semantically fits the attribute. Additionally, the human annotator adds attribute values that are not contained in the specification tables but are mentioned in the product offer to the gold standard. If an attribute is not referenced in the title or the description, the value "n/a" is assigned to this attribute.

Dataset Statistics. Table 1 shows profiling statistics describing the WDC-PAVE benchmark dataset. The dataset consists of 4,687 attribute-value pairs from 565 product which originate from 59 different websites. Overall, 45% of the attribute-value pairs hold the attribute value "n/a" meaning the attribute is neither mentioned in the title nor in the description. The dataset contains 2,011 unique attribute values so that each attribute has on average 54 unique values.

Table 1. Statistics for WDC-PAVE

Category	Home & Garden	Computers & Accessories	Grocery & Gourmet	Office Products	Jewelry	Overall
Unique Attributes	8	11	5	10	3	37
Attribute-Value Pairs	1,136	1,914	160	1,180	297	4,687
Unique Values	493	576	136	658	148	2,011
Unique Norm. Values	305	343	92	388	116	1,244
Product Offers	142	174	32	118	99	565
Host Websites	16	10	6	10	17	59

Data Normalization. Each of the 37 attribute in the dataset requires to be normalized before being usable for applications such as faceted product search. We have identified four normalization operations. Table 2 illustrates the normalization operations with examples for selected attributes. Name Expansion deals with the expansion of abbreviated attribute values such as "HP" into their non-abbreviated form, e.g. "Hewlett-Packard". Generalization assigns attribute values to broader categories, e.g. the color "Neon Lime Green" to the more general category "Green". Unit of Measurement Normalization converts an attribute value to an attribute-specific target unit of measurement and format, such as the weight value "20-lb." to "9.06", which represents the weight in kilograms (kg). String Wrangling normalizes attribute values to a specific format by for example replacing words with numbers or removing non-alphanumeric characters, e.g. the value "CTW-4M(208)" would be normalized to "CTW4M208". Each attribute is assigned to one of the normalization operations. After normalizing the 2,011 unique values in WDC-PAVE, the dataset contains 1,244 unique normalized attribute values.

Table 2. Overview of attribute value normalization operations in WDC-PAVE

Operation	Attributes	Examples
Name Expansion	Manufacturer, Generation, Capacity, Cache	"HP" → "Hewlett-Packard" "PII" → "Pentium II" "G1" → "Generation 1"
Generalization	Product Type, Color, Processor Type	"Oatmeal" → "Snacks and Breakfast" "Sparkling Juices" → "Beverages" "Neon Lime Green" → "Green"
Unit of Measurement Normalization	Dimensions, Paper Weight, Size/Weight, Rotational Speed, Pack Quantity	"7" " → "17.8" "164 ft" → "4998.7" "20-lb." → "9.06" (kg) "0.31 oz" → "879" (g) "10k" → "10000"
String Wrangling	Identifiers, Ports Processor Core, Retail UPC, Brand	"CTW-4M(208)" → "CTW4M208" "Dual Port" → "2" "4-Core" → "4" "Quaker Foods" → "QUAKER FOODS"

3 Experimental Setup

For our experiments, we split the WDC-PAVE dataset into a training set with 211 product offers and 1,750 attribute-value pairs as well as a test set with 354 product offers and 2,937 attribute-value pairs, stratified by product category. We access GPT-3.5-turbo-16k-0613 referred to as GPT-3.5 and GPT-4-0613 referred to as GPT-4 through the OpenAI API[6]. The temperature parameter of the LLMs is set to 0 to reduce the randomness. GPT-3.5 and GPT-4 are not fine-tuned. Instead, we select semantically similar demonstration product offers for in-context learning. As baselines for the experiments, we fine-tune the PLM-based extraction methods SU-OpenTag [23], AVEQA [21] and MAVEQA [26] on the training set. The fine-tuning is executed on a single NVIDIA RTX A6000 GPU. Since, the prompt templates for the LLMs and the PLM-based extraction methods utilize the same training set for demonstration selection and fine-tuning,

[6] https://platform.openai.com/docs/api-reference.

respectively, we assume that this is a fair comparison. For the evaluation, we follow related work and calculate F1-scores based on the exact match between the predicted and the ground truth attribute values [3,21,23,24,26].

4 Direct Extraction

This section compares various prompt templates for extracting attribute-value pairs from the product offers in WDC-PAVE. Following Brinkmann et al. [3], the prompts ask the LLM to extract all attributes of the target schema in a single step. Figure 2 shows an example of a complete prompt.

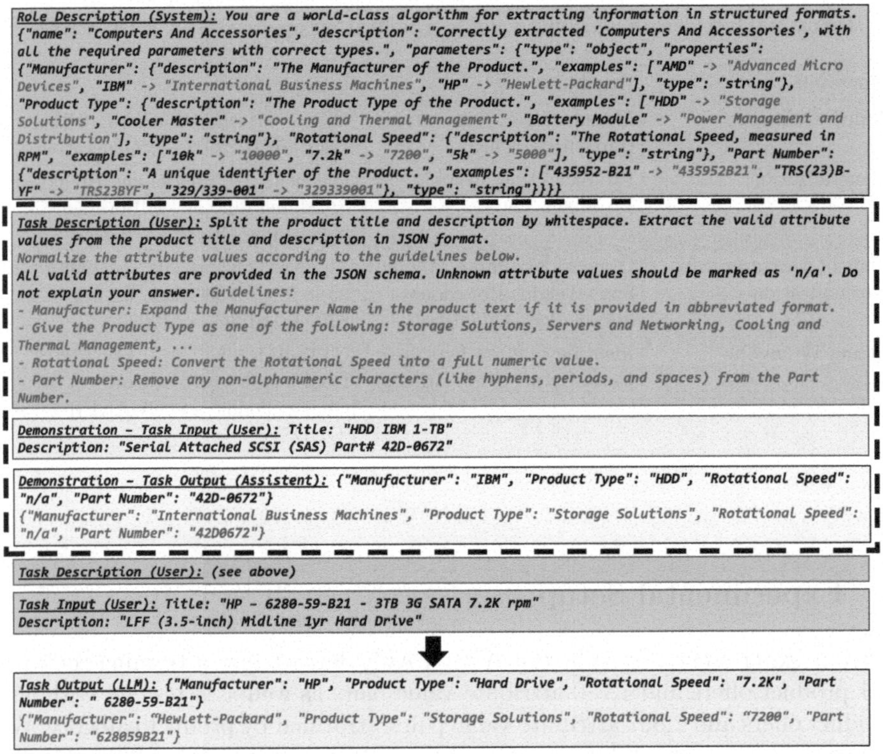

Fig. 2. Example prompt. The parts set in back font are used for the extraction. The red parts are added for the extraction with normalization task). (Color figure online)

Prompt Templates. Each prompt template consists of up to six building blocks visualized by different background colors. The building blocks are role description (blue), task description (blue), task input (green), and task output (orange), demonstration input (green) and demonstration output (yellow). The role description is a system message that defines the overall goal of the LLM as well as the target schema for the extraction including attribute descriptions and example values. The target schema is encoded using JSON-schema[7] as this representation proved to be most effective in related work [3]. The task description is a user message that provides instructions for the attribute-value extraction. The task input is a user message containing the product title and description. The task output contains the LLM's response with the extracted attribute-value pairs in JSON format. The input and output of the demonstrations are user and assistant messages containing product offers that are semantically similar to the target product offer in the task input. To calculate the semantic similarity, the target product offer and the training demonstrations are embedded using OpenAI's embedding model `text-embedding-ada-002`[8]. The training demonstrations with the highest cosine similarity to the product offer are added into the prompt. Figure 2 contains instructions for extraction and normalization. The red text is added to the prompts for the extraction with normalization task and the normalization task, which are discussed in Sect. 5 and Sect. 6.

Discussion of Results. We assess the impact of zero-shot and few-shot prompt template configurations on GPT-3.5 and GPT-4 performance. Table 3 shows the F1-scores, the average number of tokens per prompt, and the cost in $ per 1,000 extracted attribute values of the prompt template configurations with different amounts of example values (Val.) and the combination of 10 example values and different amounts of demonstrations (Dem.). The highest F1-score with and without demonstrations is marked in bold. For the cost calculation, we use the OpenAI prices as of April 2024[9]. The results show that GPT-3.5 and GPT-4 benefit from ten example values, achieving F1-scores of around 80%. Demonstrations further improve the performance of GPT-3.5 and GPT-4. Ten demonstrations allow GPT-4 to achieve the highest F1-score of 91%. Both, example values and demonstrations implicitly guide the LLMs to extract exactly the surface form of the attribute values that is used in the offer and is expected by the ground truth. The usage of example values and demonstrations significantly increases the length and the cost of the prompts. Adding three demonstrations increases the cost of extracting 1,000 attribute values with GPT-4 by 1$. In order to enhance GPT-4's performance by an additional 1.6%, an additional 2$ per 1,000 extracted attribute values must be spent.

Comparison of LLMs and PLMs. We now compare the performance of GPT-3.5 and GPT-4 with the prompt template that uses ten example values and ten demonstrations to the PLM baselines SU-OpenTag [23], AVEQA [21] and MAVEQA [26]. The baselines are fine-tuned on the same training set used to select example values and task demonstrations for the LLM prompt templates. The results in Table 4 show that GPT-4 outperforms the best PLM baseline AVEQA by 10% F1. It is important to mention that the training set with 1,750 attribute-value pairs is small. The training sets of the existing benchmark

[7] https://json-schema.org/.
[8] https://platform.openai.com/docs/guides/embeddings/.
[9] https://openai.com/pricing.

Table 3. Results for the direct extraction task.

Val.	Dem.	F1		Average length		$ per 1K values	
		GPT-3.5	GPT-4	GPT-3.5	GPT-4	GPT-3.5	GPT-4
0	0	70.66	74.40	750	745	0.09	0.10
3	0	77.11	78.77	985	973	0.12	1.25
5	0	78.51	77.52	1097	1114	0.13	1.40
10	0	80.54	79.65	1334	1338	0.15	1.66
10	3	86.91	88.94	2274	2205	0.26	2.61
10	5	86.93	88.15	2776	2735	0.32	3.21
10	10	**88.02**	**90.54**	3975	3974	0.43	4.60

datasets MAVE [26] and AE-110k [23] contain 3.8 million and 84.7 thousand attribute-value pairs. As shown by Brinkmann et al. [3], it can be expected that with additional training data the performance of the PLM baselines increases whereas the performance of the LLMs only marginally improves.

Table 4. Comparison of LLM- and PLM-based methods.

	GPT-4	GPT-3.5	AVEQA	MAVEQA	SU-OpenTag
F1	**90.54**	88.02	80.83	65.10	60.44
Δ_1 to GPT-4	-	−2.52	−9.71	−25.44	−30.10

5 Extraction with Normalization

This section evaluates prompt templates instructing LLMs to extract and normalize attribute values in a single step. To instruct the LLM on how to normalize attribute values, the prompt template in Fig. 2 is extended with the texts set in red font. The extensions add normalization guidelines to the task description and mappings between example values and their normalized forms to the target schema. The attribute values in the demonstrations are normalized.

Discussion of Results. As in Sect. 4, we measure how adding zero, three, five and ten example values without demonstrations and ten example values with three, five and ten demonstrations to the prompt affects the performance of GPT-3.5 and GPT-4. Table 5 shows the results of the experiments. The best F1-scores per model are marked in bold. Without demonstrations, GPT-4 benefits from normalized example values and reaches an F1-score of 86%, which is 12% better than the zero-shot configuration. In contrast to the extraction task, example values only marginally improve the extraction and normalization performance of GPT-3.5. However, adding five demonstrations improves the performance of both LLMs, with GPT-4 achieving an F1-score of 91% and outperforming GPT-3.5 by 5%. The prompts for the extraction with normalization task are longer than the prompts for the extraction task because of the additional normalization instructions. At the same time, the best performance is achieved with 5 instead of 10 demonstrations, resulting in a cost reduction of 1$ per 1,000 extracted attribute values for GPT-4 (see rightmost column in Table 5).

Table 5. Results for the extraction with normalization task.

Val.	Dem.	F1-score		Avg. Tok. per Prompt		$ per 1k Attr. Val.	
		GPT-3.5	GPT-4	GPT-3.5	GPT-4	GPT-3.5	GPT-4
0	0	68.62	74.19	902	889	0.1043	1.0820
3	0	71.32	83.77	1156	1155	0.1327	1.3774
5	0	70.17	84.90	1316	1312	0.1505	1.5531
10	0	68.82	85.60	1715	1673	0.1954	1.9578
10	3	85.68	91.18	2708	2702	0.3057	3.1028
10	5	**86.37**	**91.32**	3040	3079	0.3428	3.5247
10	10	86.30	91.31	4024	4031	0.4529	4.5901

Analysis of Normalization Operations. We now analyze the performance of the LLMs on the normalization operations in detail. Table 8 shows the F1-scores per normalization operation for the prompt templates zero-shot , with ten example values and with ten example values and five demonstrations. Previous research has shown that GPT-3.5 performance is weaker on tasks requiring reasoning or calculation than on tasks involving manipulation of free text or names [8]. Our results support these observations. GPT-3.5 and GPT-4 are particularly strong at name expansion and string wrangling. GPT-4 achieves F1-scores of 98% and 97%. Unit of measurement conversion requires calculations and is the most challenging operation for GPT-3.5 and GPT-4. Example values and demonstrations improve GPT-4's zero-shot performance by 22%, leading to an F1-score of 83.5%. Compared to the extraction task, we observe that generalizing attribute values simplifies the task for GPT-4, possibly because it can use its background knowledge to generalize the values. For attributes like 'Product Type', we observe that GPT-4 benefits from the generalization by an average of 7% across all product categories if example values and demonstrations from a training set are provided (Table 6).

6 Normalization

This section compares different prompt templates instructing LLMs to normalize attribute values which have been extracted by a separate preceding extraction step. As in Sect. 5, the prompt template in Fig. 2 is extended by the text set in

Table 6. F1-scores by normalization operation for the extraction with normalization task.

Normalization Operation	GPT-3.5			GPT-4		
	0 Val.	10 Val.	10 Val. 5 Dem.	0 Val.	10 Val.	10 Val. 5 Dem.
Name Expansion	41.61	42.15	**94.50**	48.64	93.60	**98.27**
Generalization	75.27	76.63	**82.16**	76.48	79.82	**88.56**
Unit of Measurement Norm.	51.16	47.64	**76.24**	61.76	73.86	**83.50**
String Wrangling	87.07	83.78	**93.37**	92.41	**97.37**	95.19

red font. The attribute values to be normalized are added to the task input block in addition to the product title and the description, which can be exploited as context in the normalization process.

Discussion of Results. Like in Sect. 4 and Sect. 5, we assess how zero, three, five and ten example values without demonstrations and ten example values with three, five and ten demonstrations selected from the training set impact the performance of GPT-3.5 and GPT-4. Table 7 shows the results of the experiments. The highest F1-scores are marked in bold. Similar as in the previous tasks, we observe that both example values and demonstrations improve the F1-scores of GPT-3.5 and GPT-4. The effect of adding demonstrations for GPT-3.5 is marginal while GPT-4 gains 4% F1-score.

Table 7. Results for the normalization task.

Val.	Dem.	F1		Average length		$ per 1K values	
		GPT-3.5	GPT-4	GPT-3.5	GPT-4	GPT-3.5	GPT-4
0	0	82.81	86.41	974	974	0.1120	1.1813
3	0	89.99	92.00	1242	1260	0.1420	1.5015
5	0	91.16	92.11	1378	1376	0.1573	1.6318
10	0	90.61	92.44	1732	1732	0.1969	2.0292
10	3	91.68	95.76	2941	2961	0.3321	3.3931
10	5	**90.98**	96.06	3486	3548	0.3931	4.0495
10	10	90.94	**96.21**	4795	4811	0.5395	5.4624

Analysis of Normalization Operations. We now analyze the normalization operations in detail. Table 8 shows the F1-scores per normalization operation for the prompt templates zero-shot, with ten example values and with ten example values and five demonstrations. The results show that string wrangling can be well handled by LLMs if the attribute values have already been extracted. The other normalization operations require example values and demonstrations to reach high F1-scores. Compared to the extraction with normalization task, the unit of measurement conversion results for GPT-3.5 and GPT-4 improve by 17% and 14% if the values have already been extracted. This shows how challenging the combination of extraction and normalization for the LLMs is given that unit of measurement conversions need to be performed. In contrast, the generalization operation remains as challenging as in the extraction with normalization task.

Table 8. F1-scores by normalization operation for the normalization task.

Normalization Operation	GPT-3.5			GPT-4		
	0 Val.	10 Val.	10 Val. 5 Dem.	0 Val.	10 Val.	10 Val. 5 Dem.
Name Expansion	64.47	95.18	**97.15**	84.10	**100.00**	99.12
Generalization	73.58	**80.67**	79.81	78.19	82.04	**90.72**
Unit of Measurement Norm.	84.89	**93.28**	93.15	90.28	94.07	**97.89**
String Wrangling	97.30	96.71	**97.70**	93.68	98.81	**99.32**

Comparison Across Tasks. We now analyze how the F1-scores change between (i) direct extraction, (ii) extraction with normalization and (iii) normalization. Table 9 shows the F1-scores of GPT-3.5 and GPT-4 for the zero-shot scenario and the few-shot scenario with ten example values and five demonstrations. In addition, the deltas between the extraction task and the extraction with normalization task (Δ_1) and the delta between the extraction task and the normalization task (Δ_2) are shown. The results indicate that extraction is more challenging for LLMs than normalization. Zero-shot both GPT-3.5 and GPT-4 achieve 12% higher F1-scores if no extraction is required. With example values and demonstrations, GPT-3.5 and GPT-4 reach F1-scores that are 4% and 8% higher if no extraction is required.

Table 9. Comparison of F1-scores across tasks.

Val.	Dem.	Model	Extract	Extract			
				Normalize	Δ_2(Extr.)	Normalize	Δ_1(Extr.)
0	0	GPT-3.5	70,66	68,62	−2,04	82,81	+12,15
0	0	GPT-4	74,40	74,19	−0,21	86,41	+12,01
10	5	GPT-3.5	86,93	86,37	−0,56	90,98	+4,05
10	5	GPT-4	88,15	91,32	3,17	96,06	+7,91

7 Related Work

Product Attribute Value Extraction. Early research on PAVE used domain-specific rules to extract attribute-value pairs [11,20,27] from product descriptions. The initial learning-based methods required extensive feature engineering and did not generalize to unknown attributes and values [6,14]. Recent works have adopted BiLSTM-CRF architectures [10,29] to tag attribute values in product titles. SU-OpenTag [23] builds upon OpenTag [29] by encoding both a target attribute and the product title using the PLM BERT [4]. [17,18,21,26] approach PAVE as a question-answering task, using different PLMs to encode target attribute, category, and title. The PLM-based methods SU-OpenTag [23], AVEQA [21], and MAVEQA [26] serve as baselines for the direct extraction task.

LLMs for Attribute Value Extraction. LLMs have successfully been used for information extraction in various domains [1,7,12]. In the context of PAVE recent works experiment with different prompt designs for PAVE using LLMs [3,5]. We use the prompt templates from [3] as a role model for our prompt templates. In contrast to related work [2,16,25], we do not fine-tune the LLMs, but instead rely on in-context learning via demonstrations and example values. Early work on attribute value extraction and normalization applied domain-specific normalization rules [15,19]. Jaimovitch-López et al. [8] experiment with using GPT-3.5 for attribute value normalization but present the values to be normalized without any context to the model. In contrast, we include the original titles and descriptions into the prompts in order to provide context for the normalization.

Benchmarks for Attribute Value Extraction and Normalization. The benchmarks MAVE [26] and AE-110k [23] are widely used to evaluate methods for PAVE. MAVE relies on an ensemble of five fine-tuned PLMs for determining ground truth annotates. AE-110k [23] uses values from product specification tables as ground truth. In contrast to these benchmarks, all attribute-value annotations in WDC-PAVE are manually verified. To our knowledge, OA-Mine [28] is the only other publicly available benchmark offering human-verified attribute-value annotations. MAVE, AE-110k, and OA-Mine address value extraction and do not consider value normalization. WDC-PAVE covers both tasks. [8] propose an attribute value normalization benchmark including operations such as transforming dates, units of measurement, or names. Unlike WDC-PAVE, their benchmark presents the values to be normalized without any context that can be exploited by the methods.

8 Conclusion

This paper investigated the ability of GPT-3.5 and GPT-4 to extract and normalize product attribute values from product offers. We experimented with different prompt templates that use example values and demonstrations for in-context learning. We introduced the WDC-PAVE benchmark, which features manually verified ground truth values for attribute value extraction as well as value normalization. GPT-4 achieves the best F1-score of 91% in the extraction task, surpassing the best PLM baseline by 10%, and shows similar performance for the extraction with normalization task. A compelling avenue for future research is to give LLMs access to scale-specific functions that the model can decide to invoke for normalizing values.

References

1. Agrawal, M., Hegselmann, S., Lang, H., Kim, Y., Sontag, D.: Large language models are few-shot clinical information extractors. In: Proceedings of the 2022 Conference on Empirical Methods in Natural Language Processing, pp. 1998–2022 (2022)
2. Blume, A., Zalmout, N., Ji, H., Li, X.: Generative models for product attribute extraction. In: Proceedings of the 2023 Conference on Empirical Methods in Natural Language Processing: Industry Track, pp. 575–585 (2023)
3. Brinkmann, A., Shraga, R., Bizer, C.: Product attribute value extraction using large language models. arXiv preprint arXiv:2310.12537 (2023)
4. Devlin, J., Chang, M.W., Lee, K., Toutanova, K.: BERT: pre-training of deep bidirectional transformers for language understanding. In: Proceedings of the 2019 Conference of the North American Chapter of the Association for Computational Linguistics: Human Language Technologies, pp. 4171–4186 (2019)
5. Fang, C., Li, X., Fan, Z., Xu, J., Nag, K., et al.: LLM-ensemble: optimal large language model ensemble method for e-commerce product attribute value extraction (2024). arXiv:2403.00863 [cs]
6. Ghani, R., Probst, K., Liu, Y., Krema, M., Fano, A.: Text mining for product attribute extraction. ACM SIGKDD Explorations Newsl **8**(1), 41–48 (2006)
7. Goel, A., Gueta, A., Gilon, O., Liu, C., Erell, S., et al.: LLMs accelerate annotation for medical information extraction. In: Proceedings of the 3rd Machine Learning for Health Symposium, pp. 82–100 (2023)
8. Jaimovitch-López, G., Ferri, C., Hernández-Orallo, J., Martínez-Plumed, F., Ramírez-Quintana, M.J.: Can language models automate data wrangling? Mach. Learn. **112**(6), 2053–2082 (2023)
9. Jain, M., Bhattacharya, S., Jain, H., Shaik, K., Chelliah, M.: Learning cross-task attribute-attribute similarity for multi-task attribute-value extraction. In: Proceedings of the 4th Workshop on e-Commerce and NLP, pp. 79–87 (2021)
10. Kozareva, Z., Li, Q., Zhai, K., Guo, W.: Recognizing salient entities in shopping queries. In: Proceedings of the 54th Annual Meeting of the Association for Computational Linguistics (Volume 2: Short Papers), pp. 107–111 (2016)
11. Nederstigt, L.J., Aanen, S.S., Vandic, D., Frasincar, F.: FLOPPIES: a framework for large-scale ontology population of product information from tabular data in e-commerce stores. Decis. Support Syst. **59**, 296–311 (2014)
12. Parekh, T., Hsu, I.H., Huang, K.H., Chang, K.W., Peng, N.: Geneva: benchmarking generalizability for event argument extraction with hundreds of event types and argument roles. In: Proceedings of the 61st Annual Meeting of the Association for Computational Linguistics, pp. 3664–3686 (2023)
13. Primpeli, A., Peeters, R., Bizer, C.: The WDC training dataset and gold standard for large-scale product matching. In: Companion Proceedings of The 2019 World Wide Web Conference, pp. 381–386 (2019)
14. Putthividhya, D., Hu, J.: Bootstrapped named entity recognition for product attribute extraction. In: Proceedings of the 2011 Conference on Empirical Methods in Natural Language Processing, pp. 1557–1567 (2011)
15. van Rooij, G., Sewnarain, R., Skogholt, M., van der Zaan, T., Frasincar, F., et al.: A data type-driven property alignment framework for product duplicate detection on the web. In: Proceedings of 17th International Web Information Systems Engineering Conference, pp. 380–395 (2016)
16. Roy, K., Goyal, P., Pandey, M.: Exploring generative frameworks for product attribute value extraction. Expert Syst. Appl. **243**, 122850 (2024)

17. Sabeh, K., Kacimi, M., Gamper, J.: CAVE: correcting attribute values in e-commerce profiles. In: Proceedings of the 31st ACM International Conference on Information & Knowledge Management, pp. 4965–4969 (2022)
18. Shinzato, K., Yoshinaga, N., Xia, Y., Chen, W.T.: Simple and effective knowledge-driven query expansion for QA-based product attribute extraction. In: Proceedings of the 60th Annual Meeting of the Association for Computational Linguistics, pp. 227–234 (2022)
19. Valstar, N., Frasincar, F., Brauwers, G.: APFA: Automated product feature alignment for duplicate detection. Expert Syst. Appl. **174**, 114759 (2021)
20. Vandic, D., Van Dam, J.W., Frasincar, F.: Faceted product search powered by the semantic web. Decis. Support Syst. **53**(3), 425–437 (2012)
21. Wang, Q., Yang, L., Kanagal, B., Sanghai, S., Sivakumar, D., et al.: Learning to extract attribute value from product via question answering: a multi-task approach. In: Proceedings of the 26th ACM SIGKDD International Conference on Knowledge Discovery and Data Mining, pp. 47–55 (2020)
22. Wei, J., Tay, Y., Bommasani, R., Raffel, C., Zoph, B., et al.: Emergent abilities of large language models. arXiv preprint arXiv:2206.07682 (2022)
23. Xu, H., Wang, W., Mao, X., Lan, M.: Scaling up open tagging from tens to thousands: comprehension empowered attribute value extraction from product title. In: Proceedings of the 57th Annual Meeting of the Association for Computational Linguistics, pp. 5214–5223 (2019)
24. Yan, J., Zalmout, N., Liang, Y., Grant, C., Ren, X., et al.: AdaTag: multi-attribute value extraction from product profiles with adaptive decoding. In: Proceedings of the 59th Annual Meeting of the Association for Computational Linguistics and the 11th International Joint Conference on Natural Language Processing, pp. 4694–4705 (2021)
25. Yang, L., Wang, Q., Wang, J., Quan, X., Feng, F., et al.: MixPAVE: mix-prompt tuning for few-shot product attribute value extraction. In: Findings of the Association for Computational Linguistics: ACL 2023, pp. 9978–9991 (2023)
26. Yang, L., Wang, Q., Yu, Z., Kulkarni, A., Sanghai, S., et al.: Mave: a product dataset for multi-source attribute value extraction. In: Proceedings of the 15th ACM International Conference on Web Search and Data Mining, pp. 1256–1265 (2022)
27. Zhang, L., Zhu, M., Huang, W.: A framework for an ontology-based E-commerce product information retrieval system. J. Comput. **4**(6), 436–443 (2009)
28. Zhang, X., Zhang, C., Li, X., Dong, X.L., Shang, J., et al.: OA-Mine: open-world attribute mining for e-commerce products with weak supervision. In: Proceedings of the ACM Web Conference 2022, pp. 3153–3161 (2022)
29. Zheng, G., Mukherjee, S., Dong, X.L., Li, F.: OpenTag: open attribute value extraction from product profiles. In: Proceedings of the 24th ACM SIGKDD International Conference on Knowledge Discovery and Data Mining, pp. 1049–1058 (2018)

Author Index

J. Tekli et al. (Eds.): ADBIS 2024, LNCS 14918, p. 231, 2024.
https://doi.org/10.1007/978-3-031-70626-4

The manufacturer's authorised representative in the EU is Springer Nature Customer Service Centre GmbH, Europaplatz 3, 69115 Heidelberg, Germany. If you have any concerns regarding our products, please contact ProductSafety@springernature.com

Printed and bound by CPI Group (UK) Ltd, Croydon, CR0 4YY
01/10/2026
02225861-0013